Business Math

Financial Math
MGMT 111

MacEwan College
Management Studies

McGraw-Hill/Irwin
A Division of The McGraw·Hill Companies

McGraw–Hill Primis

ISBN–10: 0–07–007510–7
ISBN–13: 978–0–39–051155–3

Text:

Basic Statistics for Business and
Economics, Second Canadian Edition
Lind–Marchal–Wathen–Waite

This book was printed on recycled paper.

Business Math

http://www.primisonline.com

Copyright ©2008 by The McGraw-Hill Companies, Inc. All rights reserved. Printed in the United States of America. Except as permitted under the United States Copyright Act of 1976, no part of this publication may be reproduced or distributed in any form or by any means, or stored in a database or retrieval system, without prior written permission of the publisher.

This McGraw-Hill Primis text may include materials submitted to McGraw-Hill for publication by the instructor of this course. The instructor is solely responsible for the editorial content of such materials.

111 0208GEN ISBN-10: 0-07-007510-7 ISBN-13: 978-0-39-051155-3

Business Math

Contents

Lind–Marchal–Wathen–Waite • *Basic Statistics for Business and Economics, Second Canadian Edition*

1. What is Statistics?	**1**
Text	1
2. Describing Data: Frequency Distributions and Graphic Presentation	**18**
Text	18
3. Describing Data: Numerical Measures	**54**
Text	54
6. The Normal Probability Distribution	**104**
Text	104

Chapter 1

What Is Statistics?

LEARNING OBJECTIVES

When you have completed this chapter, you will be able to:

1 Understand why we study statistics.

2 Explain what is meant by *descriptive statistics* and *inferential statistics*.

3 Distinguish between a *qualitative variable* and a *quantitative variable*.

4 Distinguish between a *discrete variable* and a *continuous variable*.

5 Distinguish among the *nominal, ordinal, interval,* and *ratio* levels of measurement.

6 Define the terms *mutually exclusive* and *exhaustive*.

Introduction

More than 100 years ago H. G. Wells, an English author and historian, noted that "statistical thinking will one day be as necessary for efficient citizenship as the ability to read." He made no mention of business because the Industrial Revolution was just beginning. Were he to comment on statistical thinking today, he would probably say that "statistical thinking is necessary not only for effective citizenship but also for effective decision making in various facets of business."

The late W. Edwards Deming, a noted statistician and quality-control expert, insisted that statistics education should begin before high school. He liked to tell the story of an 11-year-old who devised a quality-control chart to track the on-time performance of his school bus. Deming commented, "He's got a good start in life." We hope that this book will give you a solid foundation in statistics for your future life in marketing, management, accounting, sales, or some other area of business.

We apply statistical concepts almost daily in our lives. For example, when buying a CD, many people will listen to a few tracks of the CD in a listening booth in a music store before deciding whether to purchase the CD. As a second example, when students have to choose options in their courses, they will often sit in on a few sessions of a selection of courses to help them decide. In both cases, a decision is made on the basis of a sample, to buy or not buy the CD after listening to a portion of the CD, and to choose a course based on attending one or two classes.

Businesses face similar situations. The Kellogg Company must ensure that the mean amount of Raisin Bran in the 375 g box meets label specifications. To do so, they might set a "target" weight somewhat higher than the amount specified on the label. Each box is then weighed after it is filled. The weighing machine reports a distribution of the content weights for each hour as well as the number "kicked-out" for being under the label specification during the hour. The Quality Inspection Department also randomly selects samples from the production line and checks the quality of the product and the weight of the product in the box. If the mean product weight differs significantly from the target weight or the percent of kick-outs is too large, the process is adjusted.

On a national level, a candidate for the office of Prime Minister of Canada wants to know what percent of the voters in Ontario will support him in the upcoming election. There are several ways he could go about answering this question. He could have his staff call all those people in Ontario who plan to vote in the upcoming election and ask for whom they plan to vote. His staff could go out on a street in London, stop 10 people who look to be of voting age, and ask them for whom they plan to vote. He could select a random sample of about 2000 voters from the province, contact these voters, and, based on this sample, make an estimate of the percent who will vote for him in the upcoming election. In this text we will show you why the third choice is the best course of action.

Chapter 1

Why Study Statistics?

If you look through any college or university catalogue, you will find that statistics is required for many programs. Why is this so? What are the differences in the statistics courses taught for engineering, psychology, sociology, or the arts and those taught for business? The biggest difference is the examples used. The course content is basically the same. In business we are interested in such things as profits, hours worked, and wages. In psychology they are interested in test scores, and in engineering they may be interested in how many units are manufactured on a particular machine. However, all three are interested in what is a typical value and how much variation there is in the data. There may also be a difference in the level of mathematics required. An engineering statistics course usually requires calculus. Statistics courses for business usually teach the course at a more applied level.

So why is statistics required in so many majors? The first reason is that numerical information is everywhere. Look in the newspapers (The *Financial Post*), news magazines (*Time, Newsweek, Maclean's*), business magazines (*Business Week, Forbes*), or general interest magazines (*People*), women's magazines (*Chatelaine*) or sports magazines (*Sports Illustrated, ESPN The Magazine*), or the internet, and you will be bombarded with numerical information.

Examples of why we study statistics

Here are some examples:

- In 2002, the average Canadian woman earned $25 300 and the average Canadian man earned $38 900. (Adapted from the Statistics Canada CANSIM database, http://cansim2.statcan.ca, Table 202-0102, February 28, 2005.)
- Based on information collected by Statistics Canada, the most popular sport in Canada is golf (7.4%) and the least popular sport is racquetball (0.2%); the most popular sport for Canadian men is hockey (ice) at 12% and the most popular sport for Canadian women is swimming at 5.6%. (Adapted from the Statistics Canada Web site http://www.statcan.ca/english/pgdb/arts16.htm, March 21, 2005.)
- The percentage of Canadian smokers aged 15–19 years is 18.3% and aged 20–34 is 26.1%. (Adapted from Statistics Canada, Canadian Statistics, CANSIM Table 105-0027, and Catalogue no. 82-221-XIE.)
- In Canada, people drink an average of 1.15 cups of coffee per person per day. People in the United States drink more coffee than in any other country, an average of 1.75 cups per person per day.

How are we to determine if the conclusions reported are reasonable? Was the sample large enough? How were the sampled units selected? To be an educated consumer of this information, we need to be able to read the charts and graphs and understand the discussion of the numerical information. An understanding of the concepts of basic statistics will be a big help.

The second reason for taking a statistics course is that statistical techniques are used to make decisions that affect our daily lives. That is, they affect our personal welfare. Here are two examples:

- Insurance companies use statistical analysis to set rates for home, automobile, life, and health insurance. Tables are available that indicate that a 20-year old female in 2004 is expected to live to 86.0 years, and a 60-year old man in 2005 is expected to live to 82.7 years. (The *Toronto Star*, Saturday, March 19, 2005, Section D.) On the basis of these probabilities, life insurance premiums can be established.
- Medical researchers study the cure rates for diseases, based on the use of different drugs and different forms of treatment. For example, what is the effect of treating a certain type of knee injury surgically or with physical therapy? If you take an aspirin each day, does that reduce your risk of a heart attack?

A third reason for taking a statistics course is that the knowledge of statistical methods will help you understand why decisions are made and give you a better understanding of how they affect you.

No matter what line of work you select, you will find yourself faced with decisions where an understanding of data analysis is helpful. In order to make an informed decision, you will need to be able to:

Statistics in Action

We want to call your attention to a feature we call *Statistics in Action*. Read each one carefully to get an appreciation of the wide application of statistics in management, economics, nursing, law enforcement, sports, and other disciplines. Following is an assortment of statistical information, from surveys published in *Business Week, Forbes, Time*, and other magazines and newspapers.

- *Forbes* publishes annually a list of the richest Americans. According to 2004 data William Gates, founder of Microsoft Corporation, is the richest. His net worth is estimated at $58.1 billion.
- Canadian companies are ranked according to their after-tax profits in their most recent fiscal year, excluding extraordinary

What Is Statistics?   3

gains or losses, by The *Globe and Mail* (www.globeinvestor.com). The companies listed are the 1000 largest publicly traded corporations measured by assets. At the top of the list for 2003 is EnCana Corp., ranked 7th last year, followed by the Royal Bank of Canada, ranked 1st last year.

1. Determine whether the existing information is adequate or additional information is required.
2. Gather additional information, if it is needed, in such a way that it does not provide misleading results.
3. Summarize the information in a useful and informative manner.
4. Analyze the available information.
5. Draw conclusions and make inferences while assessing the risk of an incorrect conclusion.

The statistical methods presented in the text will provide you with a framework for the decision-making process.

In summary, there are at least three reasons for studying statistics: (1) data are everywhere, (2) statistical techniques are used to make many decisions that affect our lives, and (3) no matter what your future career, you will make decisions that involve data. An understanding of statistical methods will help you make these decisions more effectively.

What Is Meant by Statistics?

How do we define the word *statistics*? We encounter it frequently in our everyday language. It really has two meanings. In the more common usage, statistics refers to numerical information. Examples include the average starting salary of college graduates, the average number of Fords sold per month at Kistler Ford over the last year, the percentage of undergraduates attending McMaster who will attend graduate school, the number of deaths due to alcoholism last year, the change in the S&P/TSX Composite Index from yesterday to today, or the number of home runs hit by the Blue Jays during the 2005 season. In these examples statistics are values or percentages. Other examples include:

- Statistics Canada projects the population of Canada to be 34 419 800 by the year 2016. In 2003, 9.97% of the population of Canada lived in Alberta.
- The mean time waiting for technical support is 17 minutes.
- The mean length of the business cycle since 1945 is 61 months.

The above are all examples of **statistics.** A collection of numerical information is called **statistics** (plural).

We often present statistical information in a graphical form. A graph is a visual way to convey information and is often useful for capturing reader attention. For example, Chart 1–1 shows the average house prices in five Canadian cities in January 2005 as compared to the national average (The Canadian Real Estate Association, Jan 2005; www.crea.ca).

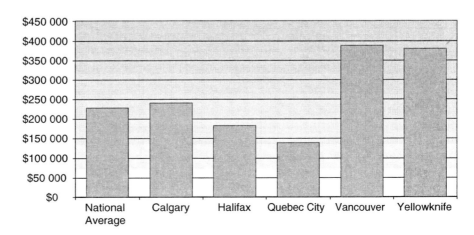

CHART 1–1 Average Canadian House Prices Jan 2005

It requires only a quick glance to see the differences, the magnitudes of which are emphasized when displayed in a bar graph.

The subject of statistics, as we will explore it in this text, has a much broader meaning than just collecting and publishing numerical information. We define statistics as:

> **STATISTICS** The science of collecting, organizing, presenting, analyzing, and interpreting data to assist in making more effective decisions.

As the definition suggests, the first step in investigating a problem is to collect relevant data. It must be organized in some way and perhaps presented in a chart, such as Chart 1–1. Only after the data have been organized are we then able to analyze and interpret it. Here are some examples of the need for data collection.

- Research analysts for Merrill Lynch evaluate many facets of a particular stock before making a "buy" or "sell" recommendation. They collect the past sales data of the company and estimate future earnings. Other factors, such as the projected worldwide demand for the company's products, the strength of the competition, and the effect of the new union-management contract, are also considered before making a recommendation.
- The marketing department at Lever Brothers, a manufacturer of soap products, has the responsibility of making recommendations regarding the potential profitability of a newly developed group of face soaps having fruit smells, such as grape, orange, and pineapple. Before making a final decision, they will test it in several markets. That is, they may advertise and sell it in Vancouver, British Columbia, and Montreal, Quebec. Based on the test marketing in these two regions, Lever Brothers will make a decision whether to market the soaps in the entire country.
- The Canadian government is concerned with the present condition of our economy and with predicting future economic trends. The government conducts a large number of surveys to determine consumer confidence and the outlook of management regarding sales and production for the next 12 months. Indexes, such as the Consumer Price Index, are constructed each month to assess inflation. Information on department store sales, housing starts, money turnover, and industrial production are just a few of the hundreds of items used to form the basis of the projections. These evaluations are used by banks to decide their prime lending rate and by the Bank of Canada to decide the level of control to place on the money supply.
- Management must make decisions on the quality of production. For example, automatic drill presses do not produce a perfect hole that is always 3.3 cm in diameter each time the hole is drilled (because of drill wear, vibration of the machine, and other factors). Slight tolerances are permitted, but when the hole is too small or too large, these products are defective and cannot be used. The Quality Assurance Department is charged with continually monitoring production by using sampling techniques to ensure that outgoing production meets standards.

Types of Statistics

Descriptive Statistics

The study of statistics is usually divided into two categories: descriptive statistics and inferential statistics. The definition of statistics given earlier referred to "organizing, presenting, analyzing . . . data." This facet of statistics is usually referred to as **descriptive statistics**.

> **DESCRIPTIVE STATISTICS** Methods of organizing, summarizing, and presenting data in an informative way.

For instance, the Canadian government reports the population of Canada was 18 238 000 in 1961, 21 568 000 in 1971, 24 820 000 in 1981, 28 031 000 in 1991, and 31 050 700 in 2001. This information is descriptive statistics. It is descriptive statistics if we

What Is Statistics?

calculate the percentage growth from one decade to the next. However, it would **not** be descriptive statistics if we used the data to forecast the population of Canada in the year 2010 or the percentage growth from 2000 to 2010.

Masses of unorganized data—such as the census of population, the weekly earnings of thousands of computer programmers, and the individual responses of 2340 registered voters regarding their choice for Prime Minister of Canada—are of little value as is. However, statistical techniques are available to organize this type of data into a meaningful form. Some data can be organized into a **frequency distribution.** (The procedure for doing this is covered in Chapter 2.) Various **charts** may be used to describe data; several basic chart forms are also presented in Chapter 2.

Specific measures of central tendency, such as the mean, describe the central value of a group of numerical data. A number of statistical measures are used to describe how closely the data cluster about an average. These measures of central location and dispersion are discussed in Chapter 3.

Inferential Statistics

Another aspect of statistics is **inferential statistics**—also called **statistical inference** and **inductive statistics.** Our main concern regarding inferential statistics is finding something about a population based on a sample taken from that population. For example, based on a sample survey by the federal government, only 46 percent of high school students can solve problems involving fractions, decimals, and percentages. And only 77 percent of high school students correctly totaled the cost of soup, a burger, fries, and a cola on a restaurant menu. Since these are inferences about the population (all high school students) based on sample data, we refer to them as inferential statistics.

> **INFERENTIAL STATISTICS** The methods used to determine something about a population, based on a sample.

Note the words *population* and *sample* in the definition of inferential statistics. We often make reference to the population living in Canada or the 1 billion population of China. However, in statistics the word *population* has a broader meaning. A *population* may consist of *individuals*—such as all the students enrolled at the University of Toronto, all the students in Accounting 201, or all the inmates of Millhaven penitentiary. A population may also consist of *objects,* such as all the XB-70 tires produced at Cooper Tire and Rubber Company, the accounts receivable at the end of October for Lorrange Plastics, Inc., or auto claims filed in the first quarter of 2005 at the Northeast Regional Office of State Farm Insurance. The *measurement* of interest might be the scores on the first examination of all students in Accounting 201, the wall weight of the Cooper tires, the dollar amount of Lorrange Plastics accounts receivable, or the amount of auto insurance claims at State Farm. Thus, a population in the statistical sense does not always refer to people.

> **POPULATION** The entire set of individuals or objects of interest or the measurements obtained from all individuals or objects of interest.

To infer something about a population, we usually take a **sample** from the population.

> **SAMPLE** A portion, or part, of the population of interest.

Reasons for sampling

Why take a sample instead of studying every member of the population? A sample of registered voters is necessary because of the prohibitive cost of contacting millions of voters before an election. Testing wheat for moisture content destroys the wheat, thus making a sample imperative. If the wine tasters tested all the wine, none would be available for sale. It would be physically impossible for a few marine biologists to capture and tag all the seals in the ocean. (These and other reasons for sampling are discussed in Chapter 7.)

Chapter 1

As noted, taking a sample to learn something about a population is done extensively in business, agriculture, politics, and government, as cited in the following examples:

- Television networks constantly monitor the popularity of their programs by hiring Nielsen and other organizations to sample the preferences of TV viewers. For example, in a sample of 800 prime-time viewers, 560 or 70 percent indicated they watched *CSI* (*Crime Scene Investigation*) last night. These program ratings are used to set advertising rates or to cancel programs.
- A public accounting firm selects a random sample of 100 invoices and checks each invoice for accuracy. There is at least one error on five of the invoices; hence the accounting firm estimates that 5 percent of the population of invoices contain at least one error.
- A random sample of 1260 marketing graduates showed their mean starting salary was $32 694. We therefore estimate the mean starting salary for all marketing graduates to be $32 694.

The relationship between a sample and a population is portrayed below. For example, we wish to estimate the mean fuel efficiency of SUVs. Six SUVs are selected from the population. The mean fuel efficiency of the six is used to estimate fuel efficiency for the population.

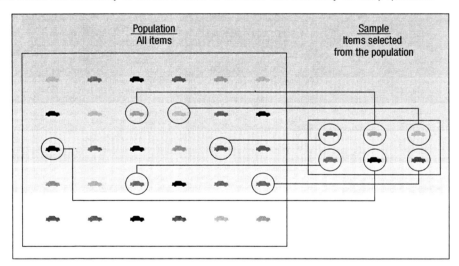

We strongly suggest you do the Self-Review exercises.

Following is a self-review problem. There are a number of them interspersed throughout each chapter. They test your comprehension of the preceding material. The answer and method of solution are given at the end of the chapter. You can find the answer to the following Self-Review at the end of the chapter. We recommend that you solve each one and then check your answer.

Self-Review 1–1

The answers are at the end of the chapter.
Morton Foods asked a sample of 1960 consumers to try a newly developed frozen fish dinner called Fish Delight. Of the 1960 sampled, 1176 said they would purchase the dinner if it were marketed.
(a) What would Morton Foods report to its Board of Directors regarding the percentage of acceptance of Fish Delight in the population?
(b) Is this an example of descriptive statistics or inferential statistics? Explain.

Types of Variables

Qualitative variable

There are two basic types of variables: (1) qualitative and (2) quantitative (see Chart 1–2). When the characteristic being studied is nonnumeric, it is called a **qualitative variable** or an

What Is Statistics?

Statistics in Action

Where did statistics get its start? In 1662 John Graunt published an article called "Natural and Political Observations Made upon Bills of Mortality." The author's "observations" were the result of his study and analysis of a weekly church publication called "Bill of Mortality," which listed births, christenings, and deaths and their causes. This analysis and interpretation of social and political data are thought to mark the start of statistics.

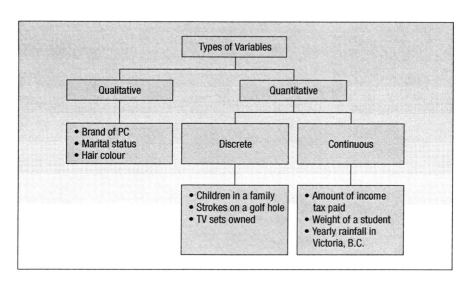

CHART 1-2 Summary of the Types of Variables

attribute. Examples of qualitative variables are gender, religious affiliation, type of automobile owned, country of birth, and eye colour. When the data are qualitative, we are usually interested in how many or what proportion fall in each category. For example, what percent of the population has blue eyes? How many Catholics and how many Protestants are there in Canada? What percent of the total number of cars sold last month were SUVs? Qualitative data are often summarized in charts and bar graphs (Chapter 2).

When the variable studied can be reported numerically, the variable is called a **quantitative variable.** Examples of quantitative variables are the balance in your chequing account, the ages of company CEOs, the life of a battery (such as 42 months), the driving speeds of automobiles, and the number of children in a family.

Quantitative variables are either discrete or continuous. **Discrete variables** can assume only certain values, and there are "gaps" between the values. Examples of discrete variables are the number of bedrooms in a house (1, 2; 3, 4, etc.), the number of cars arriving at a shopping centre in an hour (326, 421, etc.), and the number of students in each section of a statistics course (25 in section A, 42 in section B, and 18 in section C). We count, for example, the number of cars arriving at a shopping centre, and the number of statistics students in each section. Notice that a home can have 3 or 4 bedrooms, but it cannot have 3.56 bedrooms. Thus, there is a "gap" between possible values. Typically, discrete variables result from counting.

Observations of a **continuous variable** can assume any value within a specific range. Examples of continuous variables are the air pressure in a tire and the weight of a shipment of grain. Other examples are the amount of raisin bran in a box and the duration of flights from Vancouver to Calgary. Typically, continuous variables result from measuring something. On the number line, one can draw a solid line for a continuous variable because it can assume any values between two values. A discrete variable can only assume certain values on the number line, so there will be gaps between acceptable points on the number line.

Levels of Measurement

Data can be classified according to levels of measurement. The level of measurement of the data often dictates the calculations that can be done to summarize and present the data. It will also determine the statistical tests that should be performed. For example, there are six colours of candies in a bag of M&M's candies. Suppose we assign brown a value of 1, yellow 2, blue 3, orange 4, green 5, and red 6. From a bag of candies, we add the assigned colour values and divide by the number of candies and report that the mean colour is 3.56.

Does this mean that the average colour is blue or orange? As a second example, in a high school track meet there are eight competitors in the 400 m run. We report the order of finish and that the mean finish is 4.5. What does the mean finish tell us? In both of these instances, we have not properly used the level of measurement.

The levels of measurement are based on **order** and **distance**. There are actually four levels of measurement: nominal, ordinal, interval, and ratio. The lowest, or the most primitive, measurement is the nominal level. The highest, or the level that gives us the most information about the observation, is the ratio level of measurement.

Nominal Level Data

For the **nominal level** of measurement observations of a qualitative variable can only be classified and counted. There is no particular order to the labels. The classification of the six colours of M&M's candies is an example of the nominal level of measurement. We simply classify the candies by colour. There is no natural order. That is, we could report the brown candies first, the orange first, or any of the colours first. Gender is another example of the nominal level of measurement. Suppose we count the number of students entering a football game with a student ID and report how many are men and how many are women. We could report either the men or the women first. For the nominal level the only measurement involved consists of counts. Table 1–1 shows a breakdown of the sources of world oil supply. The variable of interest is the country or region. This is a nominal level variable because we record the information by country or region and there is no natural order. We could have listed the regions in alphabetical order instead of by the number of barrels produced per day. Do not be distracted by the fact that the data is summarized by reporting the number of barrels produced per day from highest to lowest.

TABLE 1–1 World Oil Supply by Country or Region

Country or Region	Millions of Barrels per Day	Percent
OPEC	28.00	37
OAPEC	19.50	26
Persian Gulf	18.84	25
United States	9.05	12
Total	75.39	100

Note: You can check the most recent information and find the countries included in the various groups by going to http://www.eia.doe.gov/emeu/ipsr/appa.html

Table 1–1 shows the essential feature of the nominal level of measurement: there is no particular order to the categories.

These categories are **mutually exclusive,** meaning, for example, that a particular barrel of oil cannot originate in more than one region at the same time.

> **MUTUALLY EXCLUSIVE** A property of a set of categories such that an individual or object is included in only one category.

The categories in Table 1–1 are also **exhaustive,** meaning that every member of the population or sample must appear in one of the categories. So the categories include all oil producing nations.

> **EXHAUSTIVE** A property of a set of categories such that each individual or object must appear in at least one category.

In order to process data on oil production, gender, employment by industry, and so forth, the categories are often numerically coded 1, 2, 3, and so on, with 1 representing OPEC, 2 representing OAPEC, for example. This facilitates counting by the computer.

What Is Statistics?

However, because we have assigned numbers to the various companies, this does not give us license to manipulate the numbers. For example, 1 + 2 does not equal 3, that is, OPEC + OAPEC does not equal Persian Gulf.

To summarize, nominal level data have the following properties:

1. Data categories are mutually exclusive and exhaustive.
2. Data categories have no logical order.

Ordinal Level Data

The next higher level of data is the **ordinal level**. Table 1–2 lists the student ratings of Professor James Brunner in an Introduction to Finance course. Each student in the class answered the question "Overall how did you rate the instructor in this class?" The variable rating illustrates the use of the ordinal scale of measurement. One classification is "higher" or "better" than the next one. That is, "Superior" is better than "Good," "Good" is better than "Average," and so on. However, we are not able to distinguish the magnitude of the differences between groups. Is the difference between "Superior" and "Good" the same as the difference between "Poor" and "Inferior"? We cannot tell. If we substitute a 5 for "Superior" and a 4 for "Good," we can conclude that the rating of "Superior" is better than the rating of "Good," but we cannot add a ranking of "Superior" and a ranking of "Good," with the result being meaningful. Further we cannot conclude that a rating of "Good" (rating is 4) is necessarily twice as high as a "Poor" (rating is 2). We can only conclude that a rating of "Good" is better than a rating of "Poor." We cannot conclude how much better the rating is.

TABLE 1–2 Rating of a Finance Professor

Rating	Frequency
Superior	6
Good	28
Average	25
Poor	12
Inferior	3

Another example of ordinal level data is found in the levels of risk of a terrorist attack to the United States. We often hear of these risks on television or read of them in newspapers. The five risk levels from lowest to highest including a description and colour codes are:

Risk Level	Description	Colour
Low	Low risk of terrorist attack	Green
Guarded	General risk of terrorist attack	Blue
Elevated	Significant risk of terrorist attack	Yellow
High	High risk of terrorist attack	Orange
Severe	Severe risk of terrorist attack	Red

This is ordinal scale data because we know the order or ranks of the risk levels—that is, orange is higher than yellow—but the amount of the difference between each of the levels is not necessarily the same. The current status is available at http://www.whitehouse.gov/homeland.

In summary, the properties of ordinal level data are:

1. The data classifications are mutually exclusive and exhaustive.
2. Data classifications are ranked or ordered according to the particular trait they possess.

Interval Level Data

The **interval level** of measurement is the next highest level. It includes all the characteristics of the ordinal level, but in addition, the difference between values is a constant size. An example of the interval level of measurement is temperature. Suppose the high temperatures on three consecutive winter days are 0, −2, and −3 degrees Celsius. These temperatures can be easily ranked, but we can also determine the difference between temperatures. This is possible because 1 degree Celsius represents a constant unit of measurement. Equal differences between two temperatures are the same, regardless of their position on the scale. That is, the difference between 10 degrees Celsius and 15 degrees is 5, the difference between 50 and 55 degrees is also 5 degrees. It is also important to note that 0 is just a point on the scale. It does not represent the absence of the condition. Zero degrees Celsius does not represent the absence of heat, just that it is cold! In fact, 0 degrees Celsius is 32 degrees Fahrenheit, and 0 degrees Fahrenheit is about −18 degrees on the Celsius scale.

The properties of interval level data are:

1. Data classifications are mutually exclusive and exhaustive.
2. Data classifications are ordered according to the amount of the characteristic they possess.
3. Equal differences in the characteristic are represented by equal differences in the measurements.

There are few examples of the interval scale of measurement. Temperature, which was just cited, is one example. Others are shoe size and IQ scores.

Ratio Level Data

Practically all quantitative data are at the ratio level of measurement. The **ratio level** is the "highest" level of measurement. It has all the characteristics of the interval level, but in addition, the 0 point is meaningful and the ratio between two numbers is meaningful. Examples of the ratio scale of measurement include: wages, units of production, weight, changes in stock prices, distance between branch offices, and height. Money is a good illustration. If you have zero dollars, then you have no money. Weight is another example. If the dial on the scale is at zero, then there is a complete absence of weight. The ratio of two numbers is also meaningful. If Jim earns $30 000 per year selling insurance and Rob earns $60 000 per year selling cars, then Rob earns twice as much as Jim.

The difference between interval and ratio measurements can be confusing. The fundamental difference involves the definition of a true zero and the ratio between two values. If you have $50 and your friend has $100, then your friend has twice as much money as you. You may convert this money to Japanese yen or English pounds, but your friend will still have twice as much money as you. If you spend your $50, then you have no money. This is an example of a true zero. As another example, a sales representative travels 150 km on Monday and 300 km on Tuesday. The ratio of the distances traveled on the two days is 2/1; converting the distances to metres or miles will not change the ratio. It is still 2/1. Suppose on Wednesday the sales representative works at home and does not travel. The distance traveled on Wednesday is zero, and this is a meaningful value. Hence, the variable distance has a true zero point.

In summary, the properties of ratio level data are:

1. Data classifications are mutually exclusive and exhaustive.
2. Data classifications are ordered according to the amount of the characteristic they possess.
3. Equal differences in the characteristic are represented by equal differences in the numbers assigned to the classifications.
4. The zero point is the absence of the characteristic.

Table 1–3 illustrates the use of the ratio scale of measurement. It shows the incomes of four father and son combinations.

TABLE 1–3 Father–Son Income Combinations

Name	Father	Son
Lahey	$80 000	$ 40 000
Nale	90 000	30 000
Rho	60 000	120 000
Steele	75 000	130 000

What Is Statistics?

Observe that the senior Lahey earns twice as much as his son. In the Rho family the son makes twice as much as the father.

Chart 1–3 summarizes the major characteristics of the various levels of measurement.

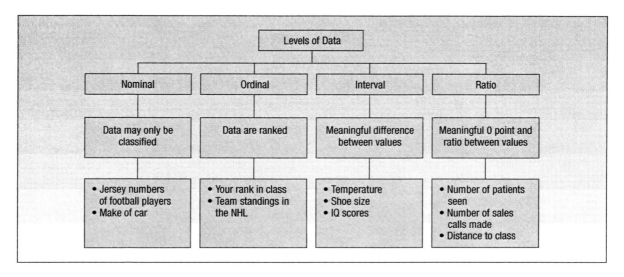

CHART 1–3 Summary of the Characteristics for Levels of Measurement

Self-Review 1–2

What is the level of measurement reflected by the following data?

(a) The ages for a sample of 50 adults who listen to the Oldies radio stations in Canada are:

35	29	41	34	44	46	42	42	37	47
30	36	41	39	44	39	43	43	44	40
47	37	41	27	33	33	39	38	43	22
44	39	35	35	41	42	37	42	38	43
35	37	38	43	40	48	42	31	51	34

(b) In a survey of 200 luxury-car owners, 100 were from Victoria, 50 from Halifax, 30 from Calgary, and 20 from Winnepeg.

Exercises

The answers to the odd-numbered exercises are at the end of the book.

1. What is the level of measurement for each of the following variables?
    a. Student IQ ratings.
    b. Distances students travel to class.
    c. Student scores on the first statistics test.
    d. A classification of students by province of birth.
    e. A ranking of students by letter grades.
    f. Numbers of hours students study per week.
2. What is the level of measurement for these items related to the newspaper business?
    a. The number of papers sold each Sunday during 2002.
    b. The number of employees in each of the departments, such as editorial, advertising, sports, etc.
    c. A summary of the number of papers sold by county.
    d. The number of years with the paper for each employee.

3. Look in the latest edition of your local newspaper and find examples of each level of measurement. Write a brief memo summarizing your findings.
4. For each of the following, determine whether the group is a sample or a population.
   a. The participants in a study of a new diabetes drug.
   b. The drivers who received a speeding ticket in Halifax last month.
   c. Those on welfare in Victoria, B.C.
   d. The 30 stocks reported as a part of the Dow Jones Industrial Average.

Uses and Abuses of Statistics

You have probably heard the old saying that there are three kinds of lies: lies, damn lies, and statistics. This saying is attributable to Benjamin Disraeli and is over a century old. It has also been said that "figures don't lie: liars figure." Both of these statements refer to the abuses of statistics in which data are presented in ways that are misleading. Many abusers of statistics are simply ignorant or careless, while others have an objective to mislead the reader by emphasizing data that support their position while leaving out data that may be detrimental to their position. One of our major goals in this text is to make you a more critical consumer of information. When you see charts or data in a newspaper, in a magazine, or on TV, always ask yourself: What is the person trying to tell me? Does that person have an agenda? Following are several examples of the abuses of statistical analysis.

An average may not be representative of all the data.

The term *average* refers to several different measures of central tendency that we discuss in Chapter 3. To most people, an average is found by adding the values involved and dividing by the number of values. So if a real estate developer tells a client that the average home in a particular subdivision sold for $250 000, we assume that $250 000 is a representative selling price for all the homes. But suppose there are only five homes in the subdivision and they sold for $150 000, $150 000, $160 000, $190 000 and $600 000. We can correctly claim that the average selling price is $250 000, but does $250 000 really seem like a "typical" selling price? Would you like to also know that the same number of homes sold for more than $160 000 as less than $160 000? Or that $150 000 is the selling price that occurred most frequently? So what selling price really is the most "typical"? This example illustrates that a reported average can be misleading, because it can be one of several numbers that could be used to represent the data. There is really no objective set of criteria that states which average should be reported on each occasion. We want to educate you as a consumer of data about how a person or group might report one value that favours their position and exclude other values. We will discuss averages, or measures of location, in Chapter 3.

Charts and graphs can also be used to visually mislead. Suppose that the price of a townhouse in Collingwood increased from $100 000 in 1995 to $200 000 in the year 2005 (see Chart 1–4). That is, the price doubled during the 10-year period. To show this change, the dollar sign on the right is twice as tall as the one on the left. However, it is also twice as wide! Therefore the area of the dollar sign on the right is 4 times (not twice) that on the left. Chart 1–4 is misleading because visually the increase is much larger than it really is.

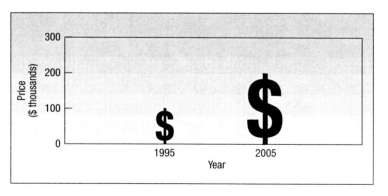

CHART 1–4 The Price of a Townhouse in Collingwood

What Is Statistics?

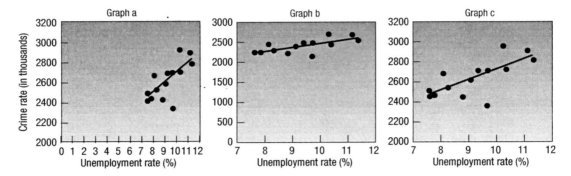

CHART 1–5 Unemployment Rate and Crime Rate in Canada

Chart 1–5 is designed to show a relationship between unemployment rate (in percent) and crime rate (in thousands, per year) in Canada in three different ways based on the same data. In Chart 1–5a, we have broken the vertical axis at 2000, and thus show a strong relation between unemployment rate and crime. In Chart 1–5b, we have broken the horizontal axis at a 7-percent rate of unemployment. In this graph, we get an impression of a weaker relation between unemployment rate and crime. A more accurate depiction of the relationship can be obtained by using values near the minimum values of the variables as starting points on each axis. Thus, a break on the vertical axis at 2000 and on the horizontal axis at 7 percent will give you a more accurate picture of the relationship as shown in Chart 1–5c.

There are many graphing techniques, but there are no hard and fast rules about drawing a graph. It is therefore both a science and an art. Your aim should always be a truthful representation of the data. The objectives and the assumptions underlying the data must be kept in mind and mentioned briefly along with graphs. The visual impressions conveyed by the graphs must correspond to the underlying data. The graphs should reveal as much information as possible with the greatest precision and accuracy. *Graphical excellence is achieved when a viewer can get the most accurate and comprehensive picture of the situation underlying the data set in the shortest possible time.* In brief, a graph should act like a mirror between the numerical data and the viewer. According to a popular saying, *"Numbers speak for themselves."* This is true for small data sets. For large data sets, it may be difficult to discern any patterns by looking at numbers alone. We therefore *need accurate portrayal of data through graphs that can speak for numbers,* and can give a quick overview of the data. We discuss graphic techniques in subsequent chapters.

Study the sampling methods.

Several years ago, a series of TV advertisements reported that "2 out of 3 dentists surveyed indicated they would recommend Brand X toothpaste to their patients." The implication is that 67 percent of all dentists would recommend the product to their patients. The trick is that the manufacturer of the toothpaste could take *many* surveys of 3 dentists and report *only* the survey of 3 dentists that had 2 indicating they would recommend Brand X. Undoubtedly, a survey of more than 3 is needed, and it must be unbiased and representative of the population of all dentists. We discuss sampling methods in Chapter 7.

Another area where there can be a misrepresentation of data is the association between variables. In statistical analysis often we find there is a strong *association* between variables. We find there is a strong association between the number of hours a student studies for an exam and the score he or she receives. Does this mean that studying causes the higher score? No. It means the two variables are related, that is, they tend to act together in a predictable fashion. We study the association between variables in Chapters 12 and 13.

Sometimes numbers themselves can be deceptive. The mean price of homes sold last month in Winnipeg, Manitoba is $129 929. This sounds like a very precise value and may

Chapter 1

instill a high degree of confidence n its accuracy. To report that the mean selling price is $130 000 doesn't convey the same precision and accuracy. However, a statistic that is very precise and carries 5 or even 10 decimal places is not necessarily accurate.

There are many other ways that statistical information can be deceiving. Entire books have been written about the subject. The most famous of these is *How to Lie with Statistics* by Darrell Huff. Understanding these practices will make you a better consumer of statistical information and help you defend yourself against those who might wish to mislead.

Computer Applications

Computers are now available to students at most colleges and universities. Spreadsheets, such as Microsoft Excel, and statistical software packages, such as MINITAB, are available in most computer labs. In this text we use Excel for the applications. We also use an Excel add-in called MegaStat. This add-in gives Excel the capability to produce additional statistical reports. Minitab examples and solutions can be found on the CD-ROM.

The following example shows the application of computers in statistical analysis. In Chapters 2 and 3 we illustrate methods for summarizing and describing data. An exercise used in those chapters refers to the list price of homes in Regina and surrounding area. The following Excel output reveals, among other things, that (1) 90 homes were listed, that (2) the mean (average) list price was $192 712.22, and (3) the list prices ranged from $62 500 to $418 900.

EXCEL

Had we used a calculator to arrive at these measures and others needed to fully analyze the list prices, hours of calculations would have been required. The likelihood of an error in arithmetic is high when a large number of values are concerned. On the other hand, statistical software packages and spreadsheets can provide accurate information in seconds.

Chapter Outline

I. Statistics is the science of collecting, organizing, presenting, analyzing, and interpreting data to assist in making more effective decisions.

II. There are two types of statistics.
   A. Descriptive statistics are procedures used to organize and summarize data.

What Is Statistics?

    B. Inferential statistics involve taking a sample from a population and making estimates about a population based on the sample results.
        1. A population is an entire set of individuals or objects of interest or the measurements obtained from all individuals or objects of interest.
        2. A sample is a part of the population.

III. There are two types of variables.
    A. A qualitative variable is nonnumeric.
        1. Usually we are interested in the number or percent of the observations in each category.
        2. Qualitative data are usually summarized in graphs and bar charts.
    B. There are two types of quantitative variables and they are usually reported numerically.
        1. Discrete variables can assume only certain values, and there are gaps between values.
        2. A continuous variable can assume any value within a specified range.

IV. There are four levels of measurement.
    A. With the nominal level, the data are sorted into categories with no particular order to the categories.
        1. The categories are mutually exclusive. An individual or object appears in only one category.
        2. The categories are exhaustive. An individual or object appears in at least one of the categories.
    B. The ordinal level of measurement presumes that one classification is ranked higher than another.
    C. The interval level of measurement has the ranking characteristic of the ordinal level of measurement plus the characteristic that the distance between values is a constant size.
    D. The ratio level of measurement has all the characteristics of the interval level, plus there is a meaningful zero point and the ratio of two values is meaningful.

Chapter Exercises

5. Explain the difference between *qualitative* and *quantitative* data. Give an example of qualitative and quantitative data.

6. Explain the difference between a sample and a population.

7. List the four levels of measurement and give an example (different from those used in the book) of each level of measurement.

8. Define the term *mutually exclusive*.

9. Define the term *exhaustive*.

10. Using data from magazines, or your local newspaper, give examples of the nominal, ordinal, interval, and ratio levels of measurement.

11. A random sample of 300 executives out of 2500 employed by a large firm showed that 270 would move to another location if it meant a substantial promotion. Based on these findings, write a brief note to management regarding all executives in the firm.

12. A random sample of 500 customers was asked to test a new toothpaste. Of the 500, 400 said it was excellent, 32 thought it was fair, and the remaining customers had no opinion. Based on these sample findings, make an inference about the reaction of all customers to the new toothpaste.

13. Explain the difference between a *discrete* and a *continuous variable*. Give an example of each not included in the text.

Data Set Exercises

14. Refer to the Real Estate data, Regina & Surrounding Area, on the CD-ROM, which reports information on listed homes and townhomes, March 2005. Consider the variables list price, type, size, number of bedrooms and number of bathrooms.

www.mcgrawhill.ca/college/lind

a. Which of the variables are qualitative and which are quantitative?
b. Determine the level of measurement for each of the variables.

15. Refer to the Tuition Fees data, which reports the average undergraduate tuition fees for full-time students, by discipline, by province, from 1999–2004. Consider the variables discipline and dollars.
   a. Which of these variables are quantitative and which are qualitative?
   b. Determine the level of measurement for each of the variables.

16. Refer to the Average Earnings by Gender and Work Pattern data, which reports information on the average salary earned by men and women in Canada from 1993–2002. The variables are year, women, men and earnings ratio (%).
   a. Which of the four variables are qualitative and which are quantitative?
   b. Determine the level of measurement for each variable.

17. Refer to the International data, which reports demographic and economic information on 46 countries.
   a. Which of the variables are quantitative and which are qualitative?
   b. Determine the level of measurement for each of the variables.

18. Refer to the CREA (Canadian Real Estate Association) data on the CD-ROM, which reports information on average house prices nationally and in a selection of cities across Canada for January and March, 2004 and 2005. Consider the variables region and average house prices.
   a. Which of the variables are quantitative and which are qualitative?
   b. Determine the level of measurement for each of the variables.

Additional exercises that require you to access information at related Internet sites are available on the CD-ROM included with this text.

What Is Statistics?

Chapter 1 Answers to Self-Reviews

1–1 a. Based on the sample of 1960 consumers, we estimate that, if it is marketed, 60 percent of all consumers will purchase Fish Delight $(1176/1960) \times 100 = 60$ percent.
b. Inferential statistics, because a sample was used to draw a conclusion about how all consumers in the population would react if Fish Delight were marketed.

1–2 a. Age is a ratio scale variable. A 40-year-old is twice as old as someone 20 years old.
b. Nominal scale. We could arrange the cities in any order.

Chapter 2

LEARNING OBJECTIVES

When you have completed this chapter, you will be able to:

1 Organize data into a frequency distribution.

2 Portray a frequency distribution in a histogram, frequency polygon, and cumulative frequency polygon.

3 Develop and interpret a stem-and-leaf display.

4 Present data using such graphical techniques as line charts, bar charts, and pie charts.

Describing Data:
Frequency Distributions and Graphic Presentation

Introduction

In this chapter we will use a selection of listings from The Canadian Real Estate Association (CREA) and The Saskatchewan Real Estate Association (SREA)'s Multiple Listing Service (http://www.mls.ca) to present techniques that organize and show the variability of house prices in Regina & Surrounding Areas. The listings provide information such as the list price, house size, type (detached, townhouse, duplex etc.), the number of bathrooms and bedrooms, and the square footage. We will use these listings to develop some tables and charts that can be reviewed to see where the list prices tend to cluster, to see the variation in the list price, and to note any trends. These techniques will be useful to anyone in managing a business.

Constructing a Frequency Distribution

Recall from Chapter 1 that we refer to techniques used to describe a set of data as *descriptive statistics*. To put it another way, we use descriptive statistics to organize data in various ways to point out where the data values tend to concentrate and help distinguish the largest and the smallest values. The first procedure we use to describe a set of data is a **frequency distribution**.

> **FREQUENCY DISTRIBUTION** A grouping of data into mutually exclusive classes showing the number of observations in each.

How do we develop a frequency distribution? The first step is to tally the data into a table that shows the classes and the number of observations in each class. The steps in constructing a frequency distribution are best described using an example. Remember, our goal is to make a table that will quickly reveal the shape of the data.

Describing Data: Frequency Distributions and Graphic Presentation

EXAMPLE

In the introduction, we describe some information available from a listing of houses for sale in Regina & Surrounding Areas. We will now use these listings to develop some tables and charts to show the typical list price. Table 2–1 reports the list price of 90 of the homes. What is the typical list price? What is the highest list price? What is the lowest list price? Around what value do the list prices tend to group?

TABLE 2–1 List Prices of Homes in Regina & Surrounding Area

				highest	
295 000	339 900	339 500	99 900	418 900	184 900
68 900	134 900	124 900	95 500	87 500	109 500
84 900	128 000	89 900	105 900	114 900	62 900
144 900	87 900	139 900	99 900	89 900	189 900
89 900	154 900	359 500	279 000	89 900	199 000
123 500	127 000	127 900	224 900	139 900	159 900
215 000	155 000	244 900	299 900	229 000	135 000
159 900	149 900	145 900	133 900	399 900	415 000
139 900	139 900	215 900	189 900	139 900	114 900
264 900	134 900	229 900	164 900	189 900	214 900
237 500	369 900	149 900	124 900	209 900	199 900
233 000	214 900	174 900	226 900	269 500	209 900
274 500	204 900	62 500	211 900	236 900	207 900
329 900	203 900	259 900	184 900	324 000	199 900
379 000	325 000	195 900	87 900	298 500	199 000
		lowest			

Solution

We refer to the unorganized information in Table 2–1 as **raw data** or **ungrouped data**. With a little searching, we can find the lowest list price ($62 500) and the highest list price ($418 900), but that is about all. It is difficult to determine a typical list price. It is also difficult to visualize where the list prices tend to cluster. The raw data are more easily interpreted if organized into a frequency distribution.

The steps for organizing data into a frequency distribution.

Step 1: Decide on the number of classes. The goal is to use just enough groupings or **classes** to reveal the shape of the distribution. Some judgment is needed here. Too many classes or too few classes might not reveal the basic shape of the set of data. In the real estate problem, for example, three classes would not give much insight into the pattern of the data (see Table 2–2).

TABLE 2–2 An Example of Too Few Classes

List Price ($ thousands)	Number of Homes
0 to under 200	53
200 to under 400	35
400 to under 600	2
Total	90

A useful recipe to determine the number of classes is the "2 to the k rule." This guide suggests you select the smallest number (k) for the number of classes such that 2^k (in words, 2 raised to the power of k) is greater than the number of observations (n).

In the real estate example, there were 90 listings. So $n = 90$. If we try $k = 6$, which means we would use 6 classes, then $2^6 = 64$, somewhat less than 90. Hence, 6 is not enough classes. If we let $k = 7$, then $2^7 = 128$, which is greater than 90. So the suggested number of classes is 7.

Step 2: Determine the class interval or width. Generally the **class interval** or width should be the same for all classes. The classes all taken together must cover at least the distance from the lowest value in the raw data up to the highest value. Expressing these words in a formula:

$$i \geq \frac{H - L}{k}$$

where i is the class interval, H is the highest observed value, L is the lowest observed value, and k is the number of classes.

In the real estate example, the lowest value is $62 500 and the highest value is $418 900. If we need 7 classes, the interval should be at least ($418 900 − $62 500)/7 = $50 914. In practice this interval size is usually rounded up to some convenient number, such as a multiple of 10 or 100, and so, the value of $55 000 might readily be used and would give us exactly 7 classes. The intervals would be $55 000–$110 000, $110 000–$165 000 etc. However, common sense tells us that an interval width of $50 000 would more readily be understood by anyone trying to interpret the data, even though using an interval width of $50 000 would give us 8 classes.

Unequal class intervals present problems in graphically portraying the distribution and in doing some of the computations that we will see in later chapters. Unequal class intervals, however, may be necessary in certain situations to avoid a large number of empty, or almost empty, classes. Such is the case in Table 2–3. Canada Customs and Revenue Agency would use unequal-sized class intervals to report the adjusted gross income on individual tax returns. Using an equal-sized interval of, say, $1000, more than 1000 classes would be required to describe all the incomes. A frequency distribution with 1000 classes would be difficult to interpret. In this case the distribution is easier to understand in spite of the unequal classes. Note also that the number of income tax returns or "frequencies" is reported in thousands in this particular table. This also makes the information easier to understand.

TABLE 2–3 Adjusted Gross Income for Individuals Filing Income Tax Returns

Adjusted Gross Income ($)	Number of Returns (thousands)
Under 2000	135
2000 to under 3000	3399
3000 to under 5000	8175
5000 to under 10 000	19 740
10 000 to under 15 000	15 539
15 000 to under 25 000	14 944
25 000 to under 50 000	4451
50 000 to under 100 000	699
100 000 to under 500 000	162
500 000 to under 1 000 000	3
1 000 000 and over	1

Step 3: Set the individual class limits. State clear class limits so you can put each observation into only one category. This means you must avoid overlapping or unclear class limits. For example, classes such as $50 000–$100 000 and $100 000–$150 000 should not be used because it is not clear whether the value of $100 000 is in the first or second class. Classes stated as $50 000–$99 000 and $100 000–$149 000 are frequently used, but may also be confusing without the additional common convention of rounding all data at or above $99 500 up to the second class and data below $99 500 down to the first class. In this text we will generally use the format $50 000 to under $100 000, $100 000 to under $150 000, and so on. With this format it is clear that $99 999 goes into the first class and $100 000 into the second class.

Describing Data: Frequency Distributions and Graphic Presentation

Statistics in Action

In 1788, James Madison, John Jay, and Alexander Hamilton anonymously published a series of essays entitled *The Federalist*. These Federalist papers were an attempt to convince the people of New York that they should ratify the Constitution. In the course of history, the authorship of most of these papers became known, but 12 remained contested. Through the use of statistical analysis, and particularly the study of the frequency of the use of various words, we can now conclude that James Madison is the likely author of the 12 papers. In fact, the statistical evidence that Madison is the author is overwhelming.

Because we round the class interval to get a convenient class size, we cover a larger than necessary range. For example, 8 classes of width $50 000 in the real estate example result in a range of 8($50 000) = 400 000. The actual range is $418 900 − $62 500 = $356 400. Comparing that value to $400 000 we have an excess of $43 600. Because we need to cover only the distance $(H - L)$, it is natural to put approximately equal amounts of the excess in each of the two tails. Of course, we should also select convenient class limits. A guideline is to make the lower limit of the first class a multiple of the class interval. Sometimes this is not possible, but the lower limit should at least be rounded. So here are the classes we could use for this data.

```
$50 000 to under $100 000
100 000 to under  150 000
150 000 to under  200 000
200 000 to under  250 000
250 000 to under  300 000
300 000 to under  350 000
350 000 to under  400 000
400 000 to under  450 000
```

The first class is $50 000 to under $100 000. The lower class limit of the first class is $50 000 and the upper class limit of the first class is $99 999.99. The second class is $100 000 to under $150 000; the lower class limit is exactly $100 000 and the upper class limit is $149 999.99, and so on.

Step 4: Tally the list prices into the classes. To begin, the first list price in Table 2–1 is $295 000. It is tallied in the $250 000 to under $300 000 class. The second list price in the first column of Table 2–1 is $68 900. It is tallied in the $50 000 to under $100 000 class. The other list prices are tallied in a similar manner. When all list prices are tallied, the table would appear as:

Class ($)	Tallies
$50 000 to under $100 000	ЖТ ЖТ IIII
100 000 to under 150 000	ЖТ ЖТ ЖТ ЖТ III
150 000 to under 200 000	ЖТ ЖТ ЖТ I
200 000 to under 250 000	ЖТ ЖТ ЖТ III
250 000 to under 300 000	ЖТ III
300 000 to under 350 000	ЖТ
350 000 to under 400 000	IIII
400 000 to under 450 000	II

Step 5: Count the number of items in each class. The number of observations in each class is called the **class frequency**. In the 50 000 to under 100 000 class there are 14 observations, and in the 300 000 to under 350 000 class, there are five observations. Therefore, the class frequency in the first class is 14 and in the sixth class is 5. There are a total of 90 observations or frequencies in the set of data.

Often it is useful to express the data in thousands, or some convenient units, rather than the actual data. Table 2–4, for example, reports the list prices in thousands of dollars, rather than dollars.

Now that we have organized the data into a frequency distribution, we can summarize the pattern in the list prices of the real estate example. Observe the following:

1. The list prices ranged from about $50 000 to under $450 000.
2. The list prices are concentrated between $50 000 and $250 000. A total of 71 list prices, or 78.9 percent of list prices, are within this range.

3. The largest concentration is in the $100 000 and $150 000 range. The middle of this class is $125 000, so we say that a typical list price is $125 000.
4. Two of the list prices were $400 000 or more, and 14 homes were listed at less than $100 000.

By presenting this information, we can see a clearer picture of the distribution of list prices.

TABLE 2–4 Frequency Distribution of List Prices, Regina & Surrounding Area

List Price ($ thousands)	Frequency
50 to under 100	14
100 to under 150	23
150 to under 200	16
200 to under 250	18
250 to under 300	8
300 to under 350	5
350 to under 400	4
400 to under 450	2
Total	90

We admit that arranging the information on list prices into a frequency distribution does result in the loss of some detailed information. That is, by organizing the data into a frequency distribution, we cannot pinpoint the exact list price, such as $134 900 or $324 000. Or, we cannot tell that the actual list price for the least expensive home was $42 500 and for the most expensive $418 900. However, the lower limit of the first class and the upper limit of the largest class convey essentially the same meaning. For example, we can tell that the highest list price is more than $400 000 but less than $450 000. The advantages of condensing the data into a more understandable form more than offset this disadvantage.

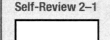

The answers are at the end of the chapter.
The commissions earned, in dollars, for the first quarter of last year by the 11 members of the sales staff at Master Chemical Company are:

| 1650 | 1475 | 1510 | 1670 | 1595 | 1760 | 1540 | 1495 | 1590 | 1625 | 1510 |

(a) What are the values such as $1650 and $1475 called?
(b) Using $1400 up to $1500 as the first class, $1500 up to $1600 as the second class, and so forth, organize the quarterly commissions into a frequency distribution.
(c) What are the numbers in the right column of your frequency distribution called?
(d) Describe the distribution of quarterly commissions, based on the frequency distribution. What is the largest amount of commission earned? What is the smallest? What is the typical amount earned?

Class Intervals and Class Midpoints

We will use two other terms frequently: **class midpoint** and **class interval.** The midpoint is halfway between the lower limits of two consecutive classes. It is computed by adding the lower limits of consecutive classes and dividing the result by 2. Referring to Table 2–4, for the first class the lower class limit is $50 000 and the next limit is $100 000. The class midpoint is $75 000, found by ($50 000 + $100 000)/2. The midpoint of $75 000 best represents, or is typical of, the list price of the homes in that class.

Describing Data: Frequency Distributions and Graphic Presentation

To determine the class interval, subtract the lower limit of the class from the lower limit of the next class. The class interval of the list price data is $50 000, which we find by subtracting the lower limit of the first class, $50 000, from the lower limit of the next class; that is, $100 000 − $50 000 = $50 000. You can also determine the class interval by finding the difference between consecutive midpoints. The midpoint of the first class is $75 000 and the midpoint of the second class is $125 000. The difference is $50 000.

A Software Example

As we mentioned in Chapter 1, there are many software packages that perform statistical calculations and output the results. Throughout this text we will show the output from Microsoft Excel and from MegaStat, which is an add-in to Microsoft Excel. MINITAB output is on the CD-ROM.

The following is a frequency distribution, produced by MegaStat, showing the list prices of the real estate data. The form of the output is somewhat different than the frequency distribution of Table 2–4, but the overall conclusions are the same.

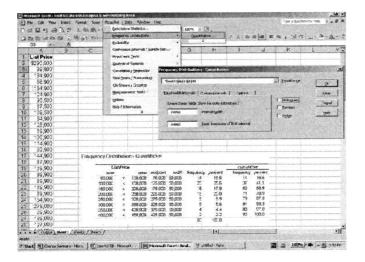

The MegaStat commands to create the frequency distribution shown above are:

1. Open Excel and the Excel file Table02-1 from the Data Files on the CD provided. From the menu bar, click MegaStat, Frequency Distributions, Quantitative.
2. In the dialogue box, input the range A1:A91; select the Equal width intervals tab; enter 50000 as the interval width and 50000 as the lower boundary of the first interval; click OK.

Self-Review 2–2

Kristina had 73 customers in her retail bicycle store, Kristina's Bikes, last Sunday. The customers spent between $48.50 and $300. Kristina wants to construct a frequency distribution of the amount spent by her customers for that day.
(a) How many classes would you use?
(b) What class interval would you suggest?
(c) What actual classes would you suggest?

Relative Frequency Distribution

A relative frequency distribution converts the frequency to a percent.

It may be desirable to convert class frequencies to **relative class frequencies** to show the fraction of the total number of observations in each class. In our real estate example, we may want to know what percent of the list prices are in the $100 000 to under $150 000 class. In another study, we may want to know what percent of the employees are absent between 1 and 3 days per year due to illness.

To convert a frequency distribution to a *relative* frequency distribution, each of the class frequencies is divided by the total number of observations. Using the distribution of list prices again (Table 2–4, where the list price is reported in thousands of dollars), the relative frequency for the $100 to under $150 class is 0.256, found by dividing 23 by 90. That is, 25.6% of the list prices are between $100 000 to under $150 000. The relative frequencies for the remaining classes are shown in Table 2–5.

TABLE 2–5 Relative Frequency Distribution of List Prices, Regina & Surrounding Area

List Price ($ thousands)	Frequency	Relative Frequency	Found by
50 to under 100	14	0.1556	14/90
100 to under 150	23	0.2556	23/90
150 to under 200	16	0.1778	16/90
200 to under 250	18	0.2000	18/90
250 to under 300	8	0.0889	8/90
300 to under 350	5	0.0556	5/90
350 to under 400	4	0.0444	4/90
400 to under 450	2	0.0222	2/90
Total	90	1.0000	

Self-Review 2–3

Refer to Table 2–5, which shows the relative frequency for the list prices of homes in Regina and surrounding area.
(a) How many homes were listed for $100 000 to under $150 000?
(b) What percent of homes were listed for $200 000 to under $250 000?
(c) What percent of the homes were listed for more than $300 000?

Exercises

The answers to the odd-numbered exercises are at the end of the book.

1. A set of data consists of 38 observations. How many classes would you recommend for the frequency distribution?
2. A set of data consists of 45 observations between $0 and $29. What size would you recommend for the class interval?
3. A set of data consists of 230 observations between $235 and $567. What class interval would you recommend?
4. A set of data contains 53 observations. The lowest value is 42 and the largest is 129. The data are to be organized into a frequency distribution.
   a. How many classes would you suggest?
   b. What would you suggest as the lower limit of the first class?
5. The Wachesaw Outpatient Center, designed for same-day minor surgery, opened last month. Following is the number of patients served for the first 16 days.

27	27	27	28	27	25	25	28
26	28	26	28	31	30	26	26

Describing Data: Frequency Distributions and Graphic Presentation

The information is to be organized into a frequency distribution.
   a. How many classes would you recommend?
   b. What class interval would you suggest?
   c. What lower limit would you recommend for the first class?
   d. Organize the information into a frequency distribution and determine the relative frequency distribution.
   e. Comment on the distribution of the values.
6. The Quick Change Oil Company has a number of outlets. The numbers of oil changes at the Oak Street outlet in the past 20 days are:

| 65 | 98 | 55 | 62 | 79 | 59 | 51 | 90 | 72 | 56 |
| 70 | 62 | 66 | 80 | 94 | 79 | 63 | 73 | 71 | 85 |

The data are to be organized into a frequency distribution.
   a. How many classes would you recommend?
   b. What class interval would you suggest?
   c. What lower limit would you recommend for the first class?
   d. Organize the number of oil changes into a frequency distribution.
   e. Comment on the distribution of the values. Also determine the relative frequency distribution.
7. The manager of the BiLo Supermarket gathered the following information on the number of times a customer visits the store during a month. The responses of 51 customers were:

5	3	3	1	4	4	5	6	4	2	6	6	6	7	1
1	14	1	2	4	4	4	5	6	3	5	3	4	5	6
8	4	7	6	5	9	11	3	12	4	7	6	5	15	1
1	10	8	9	2	12									

   a. Starting with 0 as the lower limit of the first class and using a class interval of 3, organize the data into a frequency distribution.
   b. Describe the distribution. Where do the data tend to cluster?
   c. Convert the distribution to a relative frequency distribution.
8. Moore Travel Agency, a nationwide travel agency, offers special rates on certain Caribbean cruises to senior citizens. The president of Moore Travel wants additional information on the ages of those people taking cruises. A random sample of 40 customers taking a cruise last year revealed these ages.

77	18	63	84	38	54	50	59	54	56	36	26	50	34	44
41	58	58	53	51	62	43	52	53	63	62	62	65	61	52
60	60	45	66	83	71	63	58	61	71					

   a. Organize the data into a frequency distribution, using seven classes and 15 as the lower limit of the first class. What class interval did you select?
   b. Where do the data tend to cluster?
   c. Describe the distribution.
   d. Determine the relative frequency distribution.

Graphic Presentation of a Frequency Distribution

Sales managers, stock analysts, hospital administrators, and other busy executives often need a quick picture of the trends in sales, stock prices, or hospital costs. These trends can often be depicted by the use of charts and graphs. Three charts that will help portray a frequency distribution graphically are the histogram, the frequency polygon, and the cumulative frequency polygon.

Histogram

One of the most common ways to portray a frequency distribution is a **histogram.**

> **HISTOGRAM** A bar graph where the classes are marked on the X-axis and the class frequencies on the Y-axis. The width of the bar is the class width and the height of the bar is the frequency of the class. The bars are drawn adjacent to each other.

Chapter 2

Thus, a histogram describes a frequency distribution using a series of adjacent rectangles, where the height of each rectangle is proportional to the frequency the class represents. The construction of a histogram is best illustrated by reintroducing the list prices of homes, Regina and surrounding area.

EXAMPLE

Below is the frequency distribution.

List Price ($ thousands)	Frequency
50 to under 100	14
100 to under 150	23
150 to under 200	16
200 to under 250	18
250 to under 300	8
300 to under 350	5
350 to under 400	4
400 to under 450	2
Total	90

Construct a histogram. What conclusions can you reach based on the information presented in the histogram?

Solution

The class frequencies are scaled along the vertical axis (Y-axis) and either the class limits or the class midpoints along the horizontal axis.

We note that there are 14 homes in the $50 000 to under $100 000 class. Therefore, the height of the column for that class is 14. There are 23 listings in the $100 000 to under $150 000 class, so, logically, the height of that column is 23. The height of the bar represents the number of observations in the class.

This procedure is continued for all classes. The complete histogram is shown in Chart 2–1. Note that there is no space between the bars. This is a feature of the histogram. In bar charts, which are described in a later section, the vertical bars are separated.

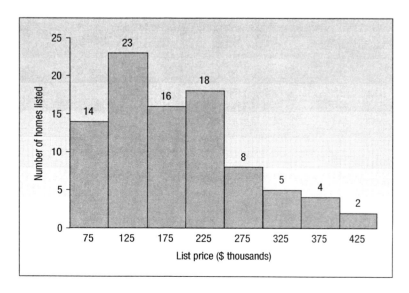

CHART 2–1 Histogram of List Prices, Regina & Surrounding Area

Describing Data: Frequency Distributions and Graphic Presentation

Based on the histogram in Chart 2–1, we conclude:

1. The lowest list price is greater than $50 000 and the highest is between $400 000 and $450 000.
2. The largest class frequency is the $100 000 to under $150 000 class. A total of 23 of the 90 listings are within this price range.
3. Seventy of the list prices, or 77.8 percent, had a list price between $100 000 and $350 000.

Thus, the histogram provides an easily interpreted visual representation of a frequency distribution. We should also point out that we would have reached the same conclusions and the shape of the histogram would have been the same had we used a relative frequency distribution instead of the actual frequencies. That is, if we had used the relative frequencies of Table 2–5, we would have had a histogram of the same shape as Chart 2–1. The only difference is that the vertical axis would have been reported in percent of listings instead of the number of listings.

We used MegaStat to produce the histogram for the real estate data. Follow the commands to create the frequency distribution. A check automatically appears in the box to the left of the word Histogram. This will produce a histogram with percents on the Y-axis. To change this default to frequencies:

1. Right click inside the plot area of the histogram. Choose Source Data.
2. Click the Series tab. You need to change the range in the Values: box.
3. In the Values: box, change the range from the Percent column to the Frequency column. Click OK. Edit the chart titles.

Frequency Polygon

In a frequency polygon the class midpoints are connected with a line segment.

A **frequency polygon** is a line graph plotting class midpoints and corresponding class frequencies. The construction of a frequency polygon is illustrated in Chart 2–2. We use the list prices for homes in Regina and surrounding area. The midpoint of each class is scaled on the X-axis and the class frequencies on the Y-axis. Recall that the class midpoint is the value at the centre of a class and represents the values in that class. The class frequency is the number of observations in a particular class. The list prices for the real estate example are:

Statistics in Action

Florence Nightingale is known as the founder of the nursing profession. However, she also saved many lives by using statistical analysis. When she encountered an unsanitary condition or an undersupplied hospital, she improved the conditions and then used statistical data to document the improvement. Thus,

(Continued on next page)

List Price ($ thousands)	Midpoint	Frequency
50 to under 100	75	14
100 to under 150	125	23
150 to under 200	175	16
200 to under 250	225	18
250 to under 300	275	8
300 to under 350	325	5
350 to under 400	375	4
400 to under 450	425	2
Total		90

As noted previously, the $50 000 to under $100 000 class is represented by the midpoint $75 000. To construct a frequency polygon, move horizontally on the graph to the midpoint, $75, and then vertically to 14, the class frequency, and place a dot. The X and the Y values of this point are called the *coordinates*. The coordinates of the next point are X = $125 and Y = 23. The process is continued for all classes. Then the points are connected in order. That is, the point representing the lowest class is joined to the one representing the second class and so on.

Chapter 2

> (Continued)
>
> she was able to convince others of the need for medical reform, particularly in the area of sanitation. She developed original graphs to demonstrate that, during the Crimean War, more soldiers died from unsanitary conditions than were killed in combat.

To produce a frequency polygon using MegaStat, follow the commands for the histogram, but select Polygon.

Note in Chart 2–2 that, to complete the frequency polygon, midpoints of $25 and $475 are added to the X-axis to "anchor" the polygon at zero frequencies. These two values, $25 and $475, were derived by subtracting the class interval of $50 from the lowest midpoint ($75) and by adding $50 to the highest midpoint ($425) in the frequency distribution.

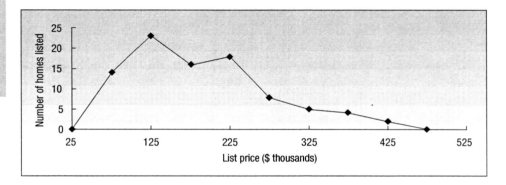

CHART 2–2 Frequency Polygon of List Prices, Regina & Surrounding Area

Both the histogram and the frequency polygon allow us to get a quick picture of the main characteristics of the data (highs, lows, points of concentration, etc.). Although the two representations are similar in purpose, the histogram has the advantage of depicting each class as a rectangle, with the height of the rectangular bar representing the number in each class. The frequency polygon, in turn, has an advantage over the histogram. It allows us to compare directly two or more frequency distributions. Suppose that the office manager at the real estate company wanted to compare list prices in the same month last year. To do this, two frequency polygons are constructed, one on top of the other, as in Chart 2–3. As expected, it is clear that the typical list price is lower for the same month in the previous year.

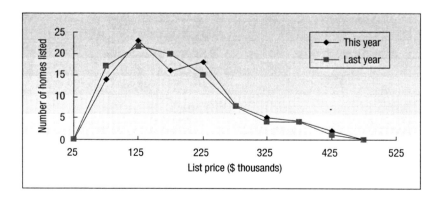

CHART 2–3 Distribution of List Prices, Regina & Surrounding Area

The total number of frequencies for the two years are about the same, so a direct comparison is possible. If the difference in the total number of frequencies is quite large, converting the frequencies to relative frequencies and then plotting the two distributions would allow a clearer comparison.

Describing Data: Frequency Distributions and Graphic Presentation

Self-Review 2–4

The annual imports of a selected group of electronics suppliers are shown in the following frequency distribution.

Imports ($ millions)	Number of Suppliers
2 to under 5	6
5 to under 8	13
8 to under 11	20
11 to under 14	10
14 to under 17	1

(a) Portray the imports as a histogram.
(b) Portray the imports as a relative frequency polygon.
(c) Summarize the important facets of the distribution (such as low and high, concentration, etc.)

Exercises

9. Molly's Candle Shop has several retail stores in coastal areas. Many of Molly's customers ask her to ship their purchases. The following chart shows the number of packages shipped per day for the last 100 days.

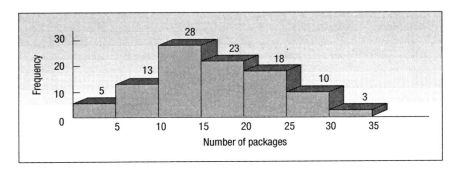

a. What is this chart called?
b. What is the total number of frequencies?
c. What is the class interval?
d. What is the class frequency for the 10 up to 15 class?
e. What is the relative frequency of the 10 up to 15 class?
f. What is the midpoint of the 10 up to 15 class?
g. On how many days were there 25 or more packages shipped?

10. The following chart shows the number of patients admitted daily to Memorial Hospital through the emergency room.

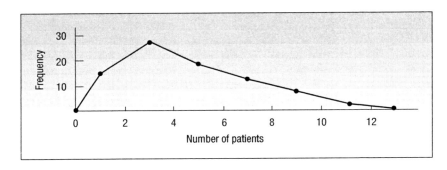

a. What is the midpoint of the 2 up to 4 class?
b. How many days were 2 up to 4 patients admitted?
c. Approximately how many days were studied?
d. What is the class interval?
e. What is this chart called?

11. The following frequency distribution reports the number of frequent flier miles, reported in thousands, for employees of Brumley Statistical Consulting, Inc. during the first quarter of 2002.

Frequent Flier Miles (thousands)	Number of Employees
0 to under 3	5
3 to under 6	12
6 to under 9	23
9 to under 12	8
12 to under 15	2
Total	50

a. How many employees were studied?
b. What is the midpoint of the first class?
c. Construct a histogram.
d. A frequency polygon is to be drawn. What are the coordinates of the plot for the first class?
e. Construct a frequency polygon.
f. Interpret the frequent flier miles accumulated using the two charts.

12. Ecommerce.com, a large Internet retailer, is studying the lead time (elapsed time between when an order is placed and when it is filled) for a sample of recent orders. The lead times are reported in days.

Lead Time (days)	Frequency
0 to under 5	6
5 to under 10	7
10 to under 15	12
15 to under 20	8
20 to under 25	7
Total	40

a. How many orders were studied?
b. What is the midpoint of the first class?
c. What are the coordinates of the first class for a frequency polygon?
d. Draw a histogram.
e. Draw a frequency polygon.
f. Interpret the lead times using the two charts.

Cumulative Frequency Distribution

Consider once again the distribution of list prices of the real estate example. Suppose we were interested in the number of homes listed for less than $150 000, or the value below which 40 percent of the homes were listed. These numbers can be approximated by developing a **cumulative frequency distribution,** and portraying it graphically in a **cumulative frequency polygon,** or **ogive.** There are two types: the less-than cumulative frequency distribution and the more-than cumulative frequency distribution. Ogives can be used to estimate percentiles such as the median (the 50th percentile).

EXAMPLE

The frequency distribution of the list prices, Regina & Surrounding Area is repeated from Table 2–4.

List Price ($ thousands)	Frequency
50 to under 100	14
100 to under 150	23
150 to under 200	16
200 to under 250	18
250 to under 300	8
300 to under 350	5
350 to under 400	4
400 to under 450	2
Total	90

Construct a less-than cumulative frequency polygon. Fifty percent of the houses were listed for less than what amount? Twenty-five of the list prices were less than what amount?

Solution

As the name implies, a cumulative frequency distribution and a cumulative frequency polygon require *cumulative frequencies*. To construct a cumulative frequency distribution, refer to the preceding table and note that there were 14 homes listed for less than $100 000. Those 14 list prices, plus the 23 in the next higher class, for a total of 37, were listed for less than $150 000. The cumulative frequency for the next higher class is 53, found by 14 + 23 + 16. This process is continued for all the classes. All the homes were listed for less than $450 000. (See Table 2–6.)

TABLE 2–6 Cumulative Frequency Distribution for List Price

List Price ($ thousands)	Frequency	Cumulative Frequency	Found by
50 to under 100	14	14	
100 to under 150	23	37	14 + 23
150 to under 200	16	53	14 + 23 + 16
200 to under 250	18	71	14 + 23 + 16 + 18
250 to under 300	8	79	
300 to under 350	5	84	
350 to under 400	4	88	
400 to under 450	2	90	
Total	90		

To plot a less-than cumulative frequency distribution, scale the upper limit of each class along the X-axis and the corresponding cumulative frequencies along the Y-axis. To provide additional information, you can label the vertical axis on the left in units and the vertical axis on the right in percent. In the real estate example, the vertical axis on the left is labeled from 0 to 90 and on the right from 0 to 100 percent. The value of 50 percent corresponds to 45 listings.

To begin plotting, 14 listings were less than $100 000, so the first point is $X = 100$ and $Y = 14$. The coordinates for the next point are $X = 150$ and $Y = 37$. The rest of the points are plotted as follows.

Chapter 2

Less than 100:	14
Less than 150:	37
Less than 200:	53
Less than 250:	71
Less than 300:	79
Less than 350:	84
Less than 400:	88
Less than 450:	90

Then the points are connected to form the chart (see Chart 2–4). To find the list price above which half the homes are listed, we draw a horizontal line from the 50 percent mark on the right-hand vertical axis over to the polygon, then drop down to the X-axis and read the list price. The value on the X-axis is about 175, so we estimate that 50 percent of the listings are less than $175 000.

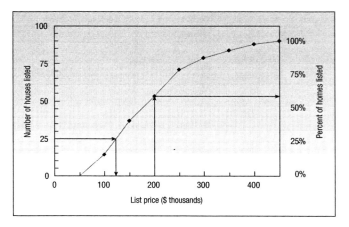

CHART 2–4 Cumulative Frequency Distribution for List Price

To find the price below which 25 of the homes are listed, we locate the value of 25 on the left-hand vertical axis. Next, we draw a horizontal line from the value of 25 to the polygon, and then drop down to the X-axis and read the price. It is about 125, so we estimate that 25 of the homes were listed for less than $125 000. We can also make estimates of the percent of homes that were listed less than a particular amount. To explain, suppose we want to estimate the percent of homes that were listed for less than $200 000. We begin by locating the value of 200 on the X-axis, move vertically to the polygon, and then horizontally to the vertical axis on the right. The value is about 60 percent, so we conclude that 60 percent of the homes were listed for less than $200 000.

To produce a less-than cumulative frequency polygon (Ogive) using MegaStat, follow the commands for the histogram, but select Ogive.

TABLE 2–7 More-than Cumulative Frequency Distribution for List Price

List Price ($ thousands)	Cumulative Frequency
More than 50:	90
More than 100:	76
More than 150:	53
More than 200:	37
More than 250:	19
More than 300:	11
More than 350:	6
More than 400:	2

To construct a more-than cumulative frequency polygon, refer to the preceding table and note that there were 90 homes listed for more than $50 000, 76 homes for more than $100 000 and so on. To plot the distribution we can use the same axes as the less-than cumulative frequency polygon. To begin the plotting, 90 listings were more than $50 000, so the first point is $X = 50$ and $Y = 90$. The coordinates for the next point are $X = 100$ and $Y = 76$. The rest of the points are plotted and then they are connected to form the chart (see Chart 2–5).

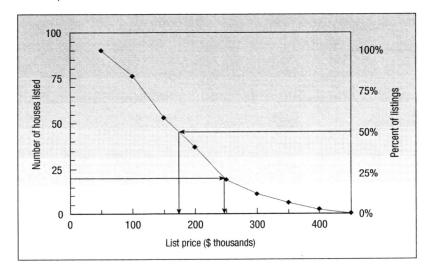

CHART 2–5 More-than Cumulative Frequency Distribution for List Price

To find the list price above which half the homes are listed, we draw a horizontal line from the 50 percent mark on the right-hand vertical axis over to the polygon, then drop down to the X-axis and read the list price. The value on the X-axis is about 175, so we estimate that 50 percent of the listings are more than $175 000. (This point is the same as for the less-than cumulative frequency polygon.) To find the price above which 20 of the homes are listed, we locate the value of 20 on the left-hand vertical axis. Next, we draw a horizontal line from the value of 20 to the polygon, and then drop down to the X-axis and read the price. It is about 250, so we estimate that 20 of the houses were listed for more than $250 000.

Self-Review 2–5

A sample of the hourly wages of 15 employees at Food City Supermarkets was organized into the following table.

Hourly Wages ($)	Number of Employees
6 to under 8	3
8 to under 10	7
10 to under 12	4
12 to under 14	1

(a) What is the table called?
(b) Develop a less-than and more-than cumulative frequency distribution and portray the distribution in cumulative frequency polygons.
(c) Based on the cumulative frequency polygon, how many employees earn $9 per hour or less? Half of the employees earn an hourly wage of how much or more? Four employees earn how much or less?

Exercises

13. The following chart shows the hourly wages of certified welders.

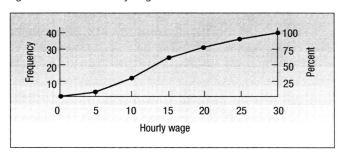

a. How many welders were studied?
b. What is the class interval?
c. About how many welders earn less than $10.00 per hour?
d. About 75 percent of the welders make less than what amount?
e. Ten of the welders studied made less than what amount?
f. What percent of the welders make less than $20.00 per hour?

14. The following chart shows the selling price, in thousands of dollars, of houses sold recently.

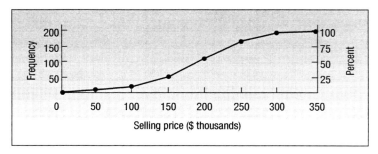

a. How many homes were studied?
b. What is the class interval?
c. One hundred homes sold for less than what amount?
d. About 75 percent of the homes sold for less than what amount?
e. Estimate the number of homes in the $150 000 up to $200 000 class.
f. About how many homes sold for less than $225 000?

15. The frequency distribution representing the number of frequent flier miles accumulated by employees at Brumley Statistical Consulting Company is repeated from Exercise 11.

Frequent Flier Miles (thousands)	Frequency
0 to under 3	5
3 to under 6	12
6 to under 9	23
9 to under 12	8
12 to under 15	2
Total	50

a. How many employees accumulated less than 3000 miles?
b. Convert the frequency distribution to a less-than and more-than cumulative frequency distribution.
c. Portray the cumulative distributions in the form of cumulative frequency polygons.
d. Based on the cumulative frequency polygons, about 75 percent of the employees accumulated how many miles or less?
e. About 40 percent of the employees accumulated more than how many miles?

16. The frequency distribution of order lead time at Ecommerce.com from Exercise 12 is repeated below.

Lead Time (days)	Frequency
0 to under 5	6
5 to under 10	7
10 to under 15	12
15 to under 20	8
20 to under 25	7
Total	40

a. How many orders were filled in less than 10 days? In less than 15 days?
b. Convert the frequency distribution to a less-than and more-than cumulative frequency distribution.
c. Develop cumulative frequency polygons.
d. About 60 percent of the orders were filled in less-than how many days?
e. How many orders were filled in more than 10 days? In more than 15 days?
f. About 25 percent of the orders were filled in more than how many days?

Stem-and-Leaf Displays

In the previous section, we showed how to organize data into a frequency distribution so we could summarize the raw data into a meaningful form. The major advantage to organizing the data into a frequency distribution is that we get a quick visual picture of the shape of the distribution without doing any further calculation. That is, we can see where the data are concentrated and also determine whether there are any extremely large or small values. There are two disadvantages, however, to organizing the data into a frequency distribution: (1) we lose the exact identity of each value and (2) we are not sure how the values within each class are distributed. To explain, the following frequency distribution shows the number of advertising spots purchased by the 45 members of the Greater Hilltown Automobile Dealers Association in the year 2000. We observe that 7 of the 45 dealers purchased between 90 and 99 spots (but less than 100). However, is the number of spots purchased within this class clustered about 90, spread evenly throughout the class, or clustered near 99? We cannot tell.

Number of Spots Purchased	Frequency
80 to under 90	2
90 to under 100	7
100 to under 110	6
110 to under 120	9
120 to under 130	8
130 to under 140	7
140 to under 150	3
150 to under 160	3
Total	45

One technique that is used to display quantitative information in a condensed form is the **stem-and-leaf display**. An advantage of the stem-and-leaf display over a frequency distribution is that we do not lose the identity of each observation. In the above example, we would not know the identity of the values in the 90 up to 100 class. To illustrate the construction of a stem-and-leaf display using the number of advertising spots purchased, suppose the seven observations in the 90 up to 100 class are: 96, 94, 93, 94, 95, 96, and 97. The **stem** value is the leading digit or digits, in this case 9. The **leaves** are the trailing digits. The stem is placed to the left of a vertical line and the leaf values to the right.

The values in the 90 up to 100 class would appear as follows:

9 | 6 4 3 4 5 6 7

Finally, we sort the values within each stem from smallest to largest. Thus, the second row of the stem-and-leaf display would appear as follows:

| 9 | 3 4 4 5 6 6 7 |

With the stem-and-leaf display, we can quickly observe that there were two dealers who purchased 94 spots and that the number of spots purchased ranged from 93 to 97. A stem-and-leaf display is similar to a frequency distribution with more information, that is, data values instead of tallies.

> **STEM-AND-LEAF DISPLAY** A statistical technique to present a set of data. Each numerical value is divided into two parts. The leading digit(s) becomes the stem and the trailing digit the leaf. The stems are located along the vertical axis, and the leaf values are stacked against each other along the horizontal axis.

The following example will explain the details of developing a stem-and-leaf display.

EXAMPLE

Listed in Table 2–8 is the number of 30-second radio advertising spots purchased by each of the 45 members of the Greater Hilltown Automobile Dealers Association last year. Organize the data into a stem-and-leaf display. Around what values do the number of advertising spots tend to cluster? What is the fewest number of spots purchased by a dealer? The largest number purchased?

TABLE 2–8 Number of Advertising Spots Purchased by Members of the Greater Hilltown Automobile Dealers Association

96	93	88	117	127	95	113	96	108	94	148	156
139	142	94	107	125	155	155	103	112	127	117	120
112	135	132	111	125	104	106	139	134	119	97	89
118	136	125	143	120	103	113	124	138			

Solution

From the data in Table 2–8 we note that the smallest number of spots purchased is 88. So we will make the first stem value 8. The largest number is 156, so we will have the stem values begin at 8 and continue to 15. The first number in Table 2–8 is 96, which will have a stem value of 9 and a leaf value of 6. Moving across the top row, the second value is 93 and the third is 88. After the first 3 data values are considered, your chart is as follows.

Stem	Leaf
8	8
9	6 3
10	
11	
12	
13	
14	
15	

Organizing all the data, the stem-and-leaf chart looks as follows.

Stem	Leaf
8	8 9
9	6 3 5 6 4 4 7
10	8 7 3 4 6 3
11	7 3 2 7 2 1 9 8 3
12	7 5 7 0 5 5 0 4
13	9 5 2 9 4 6 8
14	8 2 3
15	6 5 5

Describing Data: Frequency Distributions and Graphic Presentation

The usual procedure is to sort the leaf values from the smallest to largest. The last line, the row referring to the values in the 150s, would appear as:

15 | 5 5 6

The final table would appear as follows, where we have sorted all of the leaf values.

Stem	Leaf
8	8 9
9	3 4 4 5 6 6 7
10	3 3 4 6 7 8
11	1 2 2 3 3 7 7 8 9
12	0 0 4 5 5 5 7 7
13	2 4 5 6 8 9 9
14	2 3 8
15	5 5 6

You can draw several conclusions from the stem-and-leaf display. First the lowest number of spots purchased is 88 and the largest is 156. Two dealers purchased less than 90 spots, and three purchased 150 or more. You can observe, for example, that the three dealers who purchased more than 150 spots actually purchased 155, 155, and 156 spots. The concentration of the number of spots is between 110 and 130. There were nine dealers who purchased between 110 and 119 spots and eight who purchased between 120 and 129 spots. We can also tell that within the 120 to 129 group the actual number of spots purchased was spread evenly throughout. That is, two dealers purchased 120 spots, one dealer purchased 124 spots, three dealers purchased 125 spots, and two purchased 127 spots.

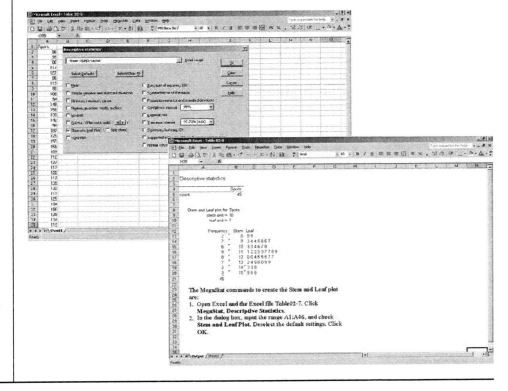

Chapter 2

Self-Review 2–6

The price-earnings ratios for 21 stocks in the retail trade category are:

8.3	9.6	9.5	9.1	8.8	11.2	7.7	10.1	9.9	10.8	
10.2	8.0	8.4	8.1	11.6	9.6	8.8	8.0	10.4	9.8	9.2

Organize this information into a stem-and-leaf display.
(a) How many values are less than 9.0?
(b) List the values in the 10.0 up to 11.0 category.
(c) What is the middle value?
(d) What are the largest and the smallest price-earnings ratios?

Exercises

17. The first row of a stem-and-leaf chart appears as follows: 62 | 1 3 3 7 9. Assume whole number values.
   a. What is the "possible range" of the values in this row?
   b. How many data values are in this row?
   c. List the actual values in this row of data.

18. The third row of a stem-and-leaf chart appears as follows: 21 | 0 1 3 5 7 9. Assume whole number values.
   a. What is the "possible range" of the values in this row?
   b. How many data values are in this row?
   c. List the actual values in this row of data.

19. The following stem-and-leaf chart shows the number of units produced per day in a factory.

3	8
4	
5	6
6	0133559
7	0236778
8	59
9	00156
10	36

   a. How many days were studied?
   b. How many observations are in the first class?
   c. What are the smallest value and the largest value?
   d. List the actual values in the fourth row.
   e. List the actual values in the second row.
   f. How many values are less than 70?
   g. How many values are 80 or more?
   h. What is the middle value?
   i. How many values are between 60 and 89, inclusive?

20. The following stem-and-leaf chart reports the number of movies rented per day at Video Connection.

12	689
13	123
14	6889
15	589
16	35
17	24568
18	268
19	13456
20	034679
21	2239
22	789
23	00179
24	8
25	13
26	
27	0

a. How many days were studied?
b. How many observations are in the last class?
c. What are the largest and the smallest values in the entire set of data?
d. List the actual values in the fourth row.
e. List the actual values in the next to the last row.
f. On how many days were less than 160 movies rented?
g. On how many days were 220 or more movies rented?
h. What is the middle value?
i. On how many days were between 170 and 210 movies rented?

21. A survey of the number of calls received by a sample of Southern Phone Company subscribers last week revealed the following information. Develop a stem-and-leaf chart. How many calls did a typical subscriber receive? What were the largest and the smallest number of calls received?

52	43	30	38	30	42	12	46	39
37	34	46	32	18	41	5		

22. Altair Banking Co. is studying the number of times their automatic teller, located in a Loblaws Supermarket, is used each day. The following is the number of times it was used during each of the last 30 days. Develop a stem-and-leaf chart. Summarize the data on the number of times the automatic teller was used: How many times was the teller used on a typical day? What were the largest and the smallest number of times the teller was used? Around what values did the number of times the teller was used tend to cluster?

83	64	84	76	84	54	75	59	70	61
63	80	84	73	68	52	65	90	52	77
95	36	78	61	59	84	95	47	87	60

Other Graphic Presentations of Data

The histogram, the frequency polygon, and the cumulative frequency polygon all have strong visual appeal. That is, they are designed to capture the attention of the reader. In this section we will examine some other graphical forms, namely the line chart, the bar chart, and the pie chart. These charts are seen extensively in newspapers, magazines, and government reports.

Line Graphs

Charts 2–6 and 2–7 are examples of **line charts**. Line charts are particularly effective for business data because we can show the change in a variable over time. The variable, such as the number of units sold or the total value of sales, is scaled along the vertical axis and time along the horizontal axis. Chart 2–6 shows the S&P/TSX Composite Index and the S&P/TSX Venture, two widely reported measures of stock market activity. Updates are available at www.tsx.com.

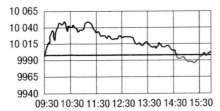
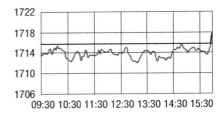

Source: TSX Datalinx, a Division of TSX Inc.

CHART 2–6 Line Charts for the S&P/TSX Composite Index and the S&P/TSX Venture Composite Index

Chapter 2

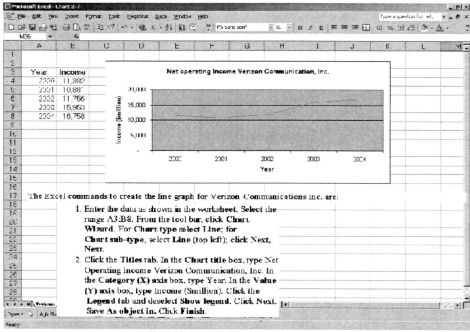

Source: TSX Datalinx, a Division of TSX Inc.

CHART 2-7 Net Operating Income for Verizon Communications, Inc. from 2000 to 2004

Chart 2–7 is also a line chart. It shows the net operating income of Verizon Communications, Inc. from 2000 to 2004. Net operating income increased from $11 392 000 in 2000 to $16 758 000 in 2004.

Quite often two or more series of data are plotted on the same chart. Thus, one chart can show the trend of those series. This allows for a comparison of several series over a period. Chart 2–8 shows the average earnings by gender for men and women from 1993 to 2002. The source of this information is Statistics Canada. We can quickly see that the average earnings for men are higher than those of women and that the earnings ratio has fluctuated somewhat from year to year, but the gap remains about the same. Note: this data set is on the CD-ROM.

Bar Charts

A bar chart can be used to depict any of the levels of measurement—nominal, ordinal, interval, or ratio. Recall we discussed the levels of data in Chapter 1. Suppose we wish to show the difference in income based on highest level of education. For example, the average income for someone over the age of 15 is $21 230 if less than a high school graduation certificate is achieved. With a high school certificate and/or some postsecondary, the income increases to $25 477. This information is summarized in Chart 2–9. (Note: The data is from Statistics Canada, Census 2001. Full details and a breakdown of income by educational level by province can be found on the CD-ROM, Data Sets.) We call Chart 2–9 a **vertical bar chart** because the bars are vertical. With this chart it is easy to see that a person with a university certificate, diploma or degree can expect to earn more than twice as much as someone with less than a high school graduation certificate. In Excel, this chart type is column.

Describing Data: Frequency Distributions and Graphic Presentation

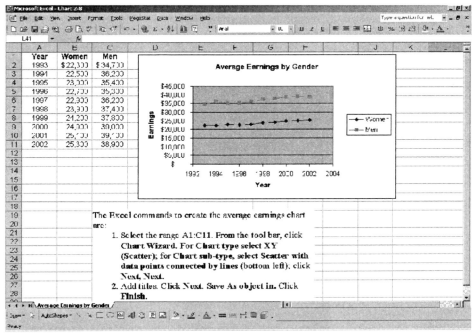

Source: Adapted from the Statistics Canada CANSIM database, http://cansim2.statcan.ca, Table 202-0102, Feb. 28, 2005.

CHART 2–8 Average Earnings by Gender from 1993 to 2002

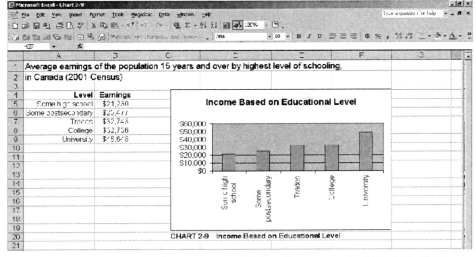

Source: Adapted from Statistics Canada, Census of Population (2001) http://www.statcan.ca/english/Pgdb/labor50a.htm; Mar 29, 2005 Data Set on the CDROM: Earnings (average) by level of schooling.

CHART 2–9 Income Based on Educational Level

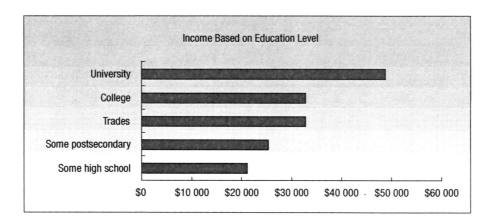

CHART 2–10 Income Based on Educational Level

Excel will also produce a horizontal bar chart. Then the chart type is bar. Chart 2–10 displays the same information in a bar chart.

There is space between the bars representing the different educational levels. This is one way in which a histogram and a bar chart differ. There is no space between the bars in a histogram (see Chart 2–1), because the list price data are ratio scale. The educational levels are nominal scale; therefore the bars are separated.

Pie Charts

A **pie chart** is especially useful for depicting nominal level data. We will use the information in Table 2–9, which shows a breakdown of lottery proceeds, to explain the details of constructing a pie chart.

TABLE 2–9 Lottery Proceeds

Use of Profits	Percent Share
Education	56
General fund	23
Cities	10
Senior citizens	9
Other	2
Total	100

The first step is to record the percentages 0, 5, 10, 15, and so on evenly around the circumference of a circle. To plot the 56 percent share for education, draw a line from 0 to the center of the circle and then another line from the center to 56 percent on the circle. The area of this "slice" represents the lottery proceeds that were given to education. Next, add the 56 percent transferred to education to the 23 percent transferred to the general fund; the result is 79 percent. Draw a line from the center of the circle to 79 percent, so the area between 56 percent and 79 percent represents the percent of the lottery proceeds transferred to the general fund. Continuing, add 10, the component given to the cities, which gives us a total of 89 percent. Draw a line from the center out to the value 89, so the area between 79 and 89 represents the share transferred to cities. Continue the same process for the senior citizen programs and "Other." Because the areas of the pie represent the relative shares of each category, we can quickly compare them: The largest percent of the proceeds goes to education; this amount is more than half the total, and it is more than twice the amount given to the next largest category.

Describing Data: Frequency Distributions and Graphic Presentation

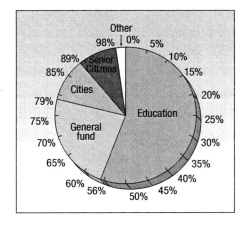

Percent of Lottery Proceeds

Following is an Excel pie chart showing the same data as in the bar and column charts, using the income based on educational levels data. We can now compare the 3 types of charts to see which chart best represents the data.

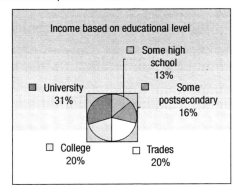

Self-Review 2–7

The Clayton City Council wants a chart to show taxpayers attending the forthcoming meeting what happens to their tax dollars. The total amount of taxes collected is $2 million. Expenditures are: $440 000 for schools, $1 160 000 for roads, $320 000 for administration, and $80 000 for supplies. A pie chart seems ideal to show the portion of each tax dollar going for schools, roads, administration, and supplies. Convert the dollar amounts to percents of the total and portray the percents in the form of a pie chart.

Exercises

23. A small business consultant is investigating the performance of several companies. The sales in 2005 (in thousands of dollars) for the selected companies were:

Corporation	Fourth-Quarter Sales ($ thousands)
Hoden Building Products	1645.2
J & R Printing, Inc.	4757.0
Long Bay Concrete Construction	8913.0
Mancell Electric and Plumbing	627.1
Maxwell Heating and Air Conditioning	24 612.0
Mizelle Roofing & Sheet Metals	191.9

The consultant wants to include a chart in his report comparing the sales of the six companies. Use a bar chart to compare the fourth quarter sales of these corporations and write a brief report summarizing the bar chart.

24. The Blair Corporation sells fashion apparel for men and women plus a broad range of home products. Listed below are the net sales for Blair from 2000 through 2005. Draw a line chart depicting the net sales over the time period and write a brief report.

Year	Net Sales ($ millions)
2000	500.0
2001	519.2
2002	526.5
2003	550.7
2004	562.9
2005	619.4

25. A headline in a newspaper reported that crime was on the decline. Listed below are the number of homicides from 1990 to 2005. Draw a line chart to summarize the data and write a brief summary of the homicide rates for the last 16 years.

Year	Homicides	Year	Homicides
1990	21	1998	40
1991	34	1999	35
1992	26	2000	30
1993	42	2001	28
1994	37	2002	25
1995	37	2003	21
1996	44	2004	19
1997	45	2005	23

26. Statistics Canada (General Social Survey, 1998) reports that of those Canadians who read books, 31.1% read at least a book a week, 36.2% at least a book a month, 17.6% at least a book every three months, 8.0% at least a book every six months, and 6.2% at least a book a year. Develop a pie chart to show the breakdown of the percentage of Canadians who read books.

27. The following table shows the population of Canada, in thousands, in five-year intervals from 1951 to 2001. Develop a line chart depicting the population growth.

Year	Population (thousands)	Year	Population (thousands)
1951	13 648	1981	24 820
1956	16 081	1986	26 101
1961	18 238	1991	28 031
1966	20 015	1996	29 672
1971	21 568	2001	31 111
1976	23 450		

28. Shown below are the student loan amounts for 2005 for eight students in the business administration program. Develop a bar chart for the data and summarize the results in a brief report.

Name	Amount ($)	Name	Amount ($)
Susan Chan	6087	Daniel Ng	3228
Sam Simone	4747	Erin Brooks	2828
Mary Suhanic	3272	Jasmine Smith	2492
Danielle Brothers	3284	Enrique Lopes	2347

Describing Data: Frequency Distributions and Graphic Presentation

Chapter Outline

I. A frequency distribution is a grouping of data into mutually exclusive classes showing the number of observations in each class.
   A. The steps in constructing a frequency distribution are:
      1. Decide how many classes you wish.
      2. Determine the class interval or width.
      3. Set the individual class limits.
      4. Tally the raw data into the classes.
      5. Count the number of tallies in each class.
   B. The class frequency is the number of observations in each class.
   C. The class interval is the difference between the limits of two consecutive classes.
   D. The class midpoint is halfway between the limits of two consecutive classes.

II. A relative frequency distribution shows the percent of the observations in each class.

III. There are three methods for graphically portraying a frequency distribution.
   A. A histogram portrays the number of frequencies in each class in the form of rectangles.
   B. A frequency polygon consists of line segments connecting the points formed by the intersections of the class midpoints and the class frequencies.
   C. A cumulative frequency polygon shows the number of observations below a certain value.

IV. A stem-and-leaf display is an alternative to a frequency distribution.
   A. The leading digits are the stem and the trailing digit the leaf.
   B. The advantages of the stem-and-leaf chart over a frequency distribution include:
      1. The identity of each observation is not lost.
      2. The digits themselves give a picture of the distribution.
      3. The cumulative frequencies are also reported.

V. There are many charts used in newspapers and magazines.
   A. A line chart is ideal for showing the trend of sales or income over time.
   B. Bar charts are similar to line charts and are useful for showing changes in nominal scale data.
   C. Pie charts are useful for showing the percent that various components are of the total.

Chapter Exercises

29. A data set consists of 83 observations. How many classes would you recommend for a frequency distribution?

30. A data set consists of 145 observations that range from 56 to 490. What size class interval would you recommend?

31. The following is the number of minutes to commute from home to work for a group of automobile executives.

28	25	48	37	41	19	32	26	16	23	23	29	36
31	26	21	32	25	31	43	35	42	38	33	28	

   a. How many classes would you recommend?
   b. What class interval would you suggest?
   c. What would you recommend as the lower limit of the first class?
   d. Organize the data into a frequency distribution.
   e. Comment on the distribution of the values.

Chapter 2

32. The following data give the weekly amounts spent on groceries for a sample of households.

271	363	159	76	227	337	295	319	250
279	205	279	266	199	177	162	232	303
192	181	321	309	246	278	50	41	335
116	100	151	240	474	297	170	188	320
429	294	570	342	279	235	434	123	325

a. How many classes would you recommend?
b. What class interval would you suggest?
c. What would you recommend as the lower limit of the first class?
d. Organize the data into a frequency distribution.

33. The following stem-and-leaf display shows the number of minutes of daytime TV viewing for a sample of college students.

0	05
1	0
2	137
3	0029
4	499
5	00155667799
6	023468
7	1366789
8	01558
9	1122379
10	022367899
11	2457
12	4668
13	249
14	5

a. How many college students were studied?
b. How many observations are in the second class?
c. What are the smallest value and the largest value?
d. List the actual values in the fourth row.
e. How many students watched less than 60 minutes of TV?
f. How many students watched 100 minutes or more of TV?
g. What is the middle value?
h. How many students watched at least 60 minutes but less than 100 minutes?

34. The following stem-and-leaf display reports the number of orders received per day by a mail-order firm.

9	1
10	2
11	235
12	69
13	2
14	135
15	1229
16	2266778
17	01599
18	00013346799
19	03346
20	4679
21	0177
22	45
23	17

a. How many days were studied?
b. How many observations are in the fourth class?
c. What are the smallest value and the largest value?
d. List the actual values in the sixth class.
e. How many days did the firm receive less than 140 orders?
f. How many days did the firm receive 200 or more orders?
g. On how many days did the firm receive 180 orders?
h. What is the middle value?

35. The following histogram shows the scores on the first statistics exam.

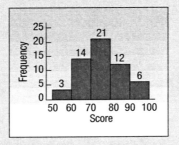

a. How many students took the exam?
b. What is the class interval?
c. What is the class midpoint for the first class?
d. How many students earned a score of less than 70?

36. The following chart summarizes the selling price of homes sold last month.

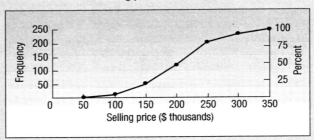

a. What is the chart called?
b. How many homes were sold during the last month?
c. What is the class interval?
d. About 75 percent of the houses sold for less than what amount?
e. One hundred seventy-five of the homes sold for less than what amount?

37. A chain of sport shops catering to beginning skiers, headquartered in Whistler, B.C., plans to conduct a study of how much a beginning skier spends on his or her initial purchase of equipment and supplies. Based on these figures, they want to explore the possibility of offering combinations, such as a pair of boots and a pair of skis, to induce customers to buy more. A sample of their cash register receipts revealed these initial purchases:

140	82	265	168	90	114	172	230	142
86	125	235	212	171	149	156	162	118
139	149	132	105	162	126	216	195	127
161	135	172	220	229	129	87	128	126
175	127	149	126	121	118	172	126	

a. Arrive at a suggested class interval. Use five classes, and let the lower limit of the first class be $80.
b. What would be a better class interval?
c. Organize the data into a frequency distribution using a lower limit of $80.
d. Interpret your findings.

Chapter 2

38. The average weekly employment insurance benefits paid in Canada for 2004 are listed below. (Statistics Canada, CANSIM table 276-0016).

	Amount ($)
Regular	293.24
Sickness	269.50
Maternity	309.24
Fishing	381.27
Work sharing	90.79
Adoption	365.57

Display the data in a vertical bar chart and a pie chart. Which chart gives a better representation of the data?

39. A recent survey showed that a typical car owner spends $2950 per year on operating expenses. Below is a breakdown of the various expenditure items. Draw an appropriate chart to portray the data and summarize your findings in a brief report.

Expenditure Item	Amount ($)
Fuel	603
Interest on car loan	279
Repairs	930
Insurance and license	646
Depreciation	492
Total	2950

40. The Lake Ontario Credit Union selected a sample of 40 student chequing accounts. Below are their end-of-the-month balances, in dollars.

404	74	234	149	279	215	123	55	43	321
87	234	68	489	57	185	141	758	72	863
703	125	350	440	37	252	27	521	302	127
968	712	503	489	327	608	358	425	303	203

a. Tally the data into a frequency distribution using $100 as a class interval and $0 as the starting point.
b. Draw a cumulative frequency polygon.
c. The bank considers any student with an ending balance of $400 or more a "preferred customer." Estimate the percentage of preferred customers.
d. The bank is also considering a service charge to the lowest 10 percent of the ending balances. What would you recommend as the cutoff point between those who have to pay a service charge and those who do not?

41. The estimate of the median age of the population in Canada, the provinces and the Yukon Territory are listed below. Construct a stem-and-leaf chart from the data. Summarize your findings. Note that more details of the median age and the population are available on the CD-ROM, Data Sets.

	Median Age (2003)
Canada	37.9
Newfoundland and Labrador	39.3
Prince Edward Island	38.6
Nova Scotia	39.5
New Brunswick	39.2
Quebec	39.4
Ontario	37.4
Manitoba	36.8
Saskatchewan	36.9
Alberta	35.1
British Columbia	38.8
Yukon Territory	36.3

Describing Data: Frequency Distributions and Graphic Presentation

42. A recent study of home technologies reported the number of hours of personal computer usage per week for a sample of 60 persons. Excluded from the study were people who worked out of their home and used the computer as a part of their work.

9.3	5.3	6.3	8.8	6.5	0.6	5.2	6.6	9.3	4.3
6.3	2.1	2.7	0.4	3.7	3.3	1.1	2.7	6.7	6.5
4.3	9.7	7.7	5.2	1.7	8.5	4.2	5.5	5.1	5.6
5.4	4.8	2.1	10.1	1.3	5.6	2.4	2.4	4.7	1.7
2.0	6.7	1.1	6.7	2.2	2.6	9.8	6.4	4.9	5.2
4.5	9.3	7.9	4.6	4.3	4.5	9.2	8.5	6.0	8.1

a. Organize the data into a frequency distribution. How many classes would you suggest? What value would you suggest for a class interval?
b. Draw a histogram. Interpret your result.

43. Merrill Lynch recently completed a study regarding the size of investment portfolios (stocks, bonds, mutual funds, and certificates of deposit) for a sample of clients in the 40 to 50 age group. Listed below is the value of all the investments in thousands of dollars for the 70 participants in the study.

669.9	7.5	77.2	7.5	125.7	516.9	219.9	645.2
301.9	235.4	716.4	145.3	26.6	187.2	315.5	89.2
136.4	616.9	440.6	408.2	34.4	296.1	185.4	526.3
380.7	3.3	363.2	51.9	52.2	107.5	82.9	63.0
228.6	308.7	126.7	430.3	82.0	227.0	321.1	403.4
39.5	124.3	118.1	23.9	352.8	156.7	276.3	23.5
31.3	301.2	35.7	154.9	174.3	100.6	236.7	171.9
221.1	43.4	212.3	243.3	315.4	5.9	1002.2	171.7
295.7	437.0	87.8	302.1	268.1	899.5		

a. Organize the data into a frequency distribution. How many classes would you suggest? What value would you suggest for a class interval?
b. Draw a histogram. Interpret your result.

44. The following table displays the average retail price, in cents per litre, of gasoline in St. John's, Newfoundland and Labrador over six years. Develop a bar chart depicting this information. Note that more details on gasoline prices in Canada are available on the CD-ROM, Data Sets.

Year	2003	2002	2001	2000	1999	1998
Cents per litre	85.9	80.1	81.7	85.6	68.9	67.0

45. The business school at the local college reported the following percentage breakdown of expenses. Draw a pie chart showing the information.

Category	Percent
Professional development	32.3
Equipment	23.5
Mileage	12.6
PAC meetings	12.1
Office supplies	10.9
Other	8.6

Chapter 2

46. In their 2003 annual report Schering-Plough Corporation reported their income, in millions of dollars, for the years 1998 to 2003 as follows. Develop a line chart depicting the results and comment on your findings. Note that there was a $46 million loss in 2003.

Year	Income ($ millions)
1998	1756
1999	2110
2000	2423
2001	1943
2002	1974
2003	(46)

47. The average earnings ($ per year) of people 15 years and over who have a college certificate or diploma are listed below by province. Develop an appropriate chart and write a brief report summarizing the information. The data is from Statistics Canada, 2001 Census. Note that more details of the average earnings of Canadians are available on the CD-ROM, Data Sets.

Canada	32 736
Newfoundland and Labrador	28 196
Prince Edward Island	25 613
Nova Scotia	26 930
New Brunswick	27 178
Quebec	28 742
Ontario	36 309
Manitoba	29 351
Saskatchewan	27 742
Alberta	33 572
British Columbia	33 159
Yukon	33 817

(Adapted from Statistics Canada, Census of Population (2001) http://www.statcan.ca/english/Pgdb/labor50a.htm; Mar 29, 2005.)

48. Annual imports from selected Canadian trading partners are listed below for the year 2003. Develop an appropriate chart or graph and write a brief report summarizing the information.

Partner	Annual Imports ($ millions)
Japan	9550
United Kingdom	4556
South Korea	2441
China	1182
Australia	618

49. A large pharmaceutical company with distribution centres in Canada and the United States has its call centre located in Mississauga, Ontario. There are twenty customer service representatives answering calls from 8 a.m. to 9 p.m. through the week, and from 9 a.m. to 5 p.m. on Saturdays. The call centre reported the number of calls waiting to be answered between the hours of 9 a.m. and 10 a.m. over a fifty-day period as shown below. Summarize the data in a chart and interpret.

47	1	8	46	76	26	4	3	39	45
4	21	80	63	100	65	91	29	7	15
7	52	87	39	106	25	55	2	3	8
14	38	59	33	76	71	37	51	1	24
35	86	185	13	7	43	36	20	79	9

Describing Data: Frequency Distributions and Graphic Presentation

50. One of the most popular candies in North America is M&M's, which is produced by the Mars Company. For many years the M&M's plain candies were produced in six colours: red, green, orange, tan, brown, and yellow. Recently, tan was replaced by blue. Did you ever wonder how many candies were in a bag, or how many of each colour? Are there about the same number of each colour, or are there more of some colours than others? Here is some information for a 500 g bag of M&M's plain candies. It contained a total of 544 candies. There were 135 brown, 156 yellow, 128 red, 22 green, 50 blue, and 53 orange. Develop a chart depicting this information and a brief report summarizing the information.

51. The following graph compares the average earnings of men and women in Canada from 1992 to 2004. Write a brief report summarizing the information in the graph. Be sure to include the changes that you see each period and the direction of the change over the twelve-year period.

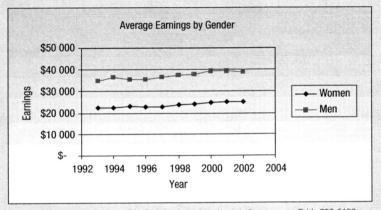

(Adapted from Statistics Canada CANSIM database, http://cansim2.statcan.ca, Table 202-0102, Feb 28, 2005.)

Data Set Exercises

52. Refer to the Real Estate data, Regina & Surrounding Area, on the CD-ROM, which reports information on listed homes and townhomes, March 2005.
    a. Select an appropriate class interval and organize the list prices into a frequency distribution.
       1. Around what values do the data tend to cluster?
       2. What is the largest list price? What is the smallest list price?
    b. Draw a less-than and more-than cumulative frequency polygon based on the frequency distribution developed in part (a).
       1. How many homes listed for less than $200 000?
       2. Estimate the percent of the homes that listed for more than $220 000.
       3. What percent of the homes listed for less than $325 000?
    c. Organize the information on the size (sq ft) into a frequency distribution. Select an appropriate class interval.
       1. What is a typical size? What is the range of sizes?
       2. Comment on the distribution of the values. Does it appear that any of the sizes are out of line with the others?
    d. Draw cumulative frequency distributions based on the frequency distributions developed in part (c).
       1. Forty percent of the sizes are less than what size?
       2. About how many houses have total square feet of less than 1200?
       3. What percent of homes are greater than 2000 square feet?
    e. Develop a stem-and-leaf chart for the list price.
       1. Around what value are the list prices clustered?
       2. Are there any outliers?
       3. Summarize the results in a brief report.
    f. Develop a stem-and-leaf chart for the size (sq ft).
       1. Around what value are the number of square feet clustered?
       2. Are there any outliers?
       3. Summarize the results in a brief report.

Chapter 2

53. Refer to the International data, which reports demographic and economic information on 46 countries.
  a. Develop a frequency distribution for the variable GNP per capita. Summarize your findings. Comment on the distribution of the values.
  b. Develop a stem-and-leaf chart for the variable referring to the number of cell phones. Summarize your findings.

Case

Rob Whitner is the owner of Whitner Pontiac. Rob's father founded the dealership in 1964, and for more than 30 years they sold exclusively Pontiacs. In the early 1990's Rob's father's health began to fail, and Rob took over more of the day-to-day operation of the dealership. At this same time, the automobile business began to change—dealers began to sell vehicles from several manufacturers—and Rob was faced with some major decisions. The first came when another local dealer, who handles Volvos, Saabs, and Volkswagens, approached Rob about purchasing his dealership. More recently, the local Chrysler dealership got into difficulty and Rob bought them out. So now, on the same lot, Rob sells the complete line of Pontiacs, the expensive Volvos, Saabs, Volkswagens, and the Chrysler products, including the popular Jeep line. Whitner Pontiac employs 83, including 23 full-time salespeople. Because of the diverse product line, there is considerable variation in the selling price of the vehicles. A top-of-the-line Volvo sells for more than twice the price of a Pontiac Grand Am. Rob would like you to develop some tables and charts that he could review monthly and would like you to report where the selling prices tend to cluster, where the variation is in the selling prices, and to note any trends. The data is on the CD-ROM, Data Files, Whitner-2005.

Additional exercises that require you to access information at related Internet sites are available on the CD-ROM included with this text.

Chapter 2 Answers to Self-Reviews

2-1 a. The raw data.

b.

Commission ($)	Number of Salespeople
1400 up to 1500	2
1500 up to 1600	5
1600 up to 1700	3
1700 up to 1800	1
Total	11

c. Class frequencies.
d. The largest concentration of commissions is $1500 up to $1600. The smallest commission is about $1400 and the largest is about $1800.

2-2 a. $2^6 = 64 < 73 < 128 = 2^7$. So 7 classes are recommended.
b. The interval width should be at least $(300 - 48.50)/7 = 35.93$. Class intervals of 30, 35, or 40 are reasonable.
c. If we use a class interval of 30, and begin with $40, 9 classes would be required. A class interval of 40, starting with $40, would require 7 classes.

2-3 a. 23
b. 22.2%, found by $(20/90) \times 100$
c. 12.2%, found by $(11/90) \times 100$

2-4 a.
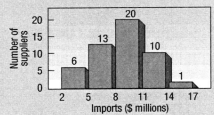

b.
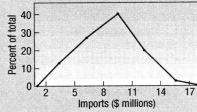

The points are: (3.5, 12), (6.5, 26), (9.5, 40), (12.5, 20), and (15.5, 2).

c. The smallest annual sales volume of imports by a supplier is about $2 million, the highest about $17 million. The concentration is between $8 million and $11 million.

2-5 a. A frequency distribution.

b.

Hourly Wages ($)	less-than Cumulative Number	more-than Cumulative Number
6 to under 8	3	15
8 to under 10	10	12
10 to under 12	14	5
12 to under 14	15	1

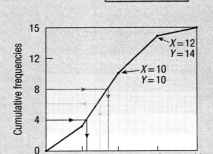

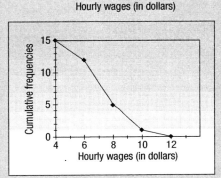

c. About seven employees earn $9.00 or less. About half the employees earn $7.25 or more. About four employees earn $8.25 or less.

2-6

7	7
8	0013488
9	1256689
10	1248
11	26

a. 8
b. 10.1, 10.2, 10.4, 10.8
c. 9.5
d. 7.7, 11.6

2-7
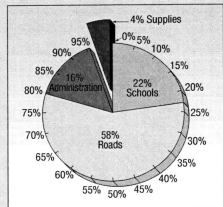

Chapter 3

Describing Data: Numerical Measures

LEARNING OBJECTIVES

When you have completed this chapter, you will be able to:

1 Calculate the *arithmetic mean, median, mode, weighted mean,* and *geometric mean.*

2 Explain the characteristics, uses, advantages, and disadvantages of each *measure of location.*

3 Identify the position of the *arithmetic mean, median,* and *mode* for both symmetric and skewed distributions.

4 Compute and interpret the *range, mean deviation, variance,* and *standard deviation.*

5 Explain the characteristics, uses, advantages, and disadvantages of each *measure of dispersion.*

6 Understand *Chebyshev's theorem* and the *Empirical Rule* as they relate to a set of observations.

7 Compute and interpret *quartiles, interquartile range,* and *percentiles.*

8 Compute and interpret the *coefficient of skewness* and the *coefficient of variation.*

Introduction

Chapter 2 began our study of descriptive statistics. To transform a mass of raw data into a meaningful form, we organized it into a frequency distribution and portrayed it graphically in a histogram or a frequency polygon. We also looked at other graphical techniques such as line charts and pie charts.

This chapter is concerned with two numerical ways of describing data, namely, **measures of location** and **measures of dispersion.** Measures of location are often referred to as **averages.** The purpose of a measure of central location is to pinpoint the centre of a set of values.

You are familiar with the concept of an average. Averages appear daily on TV, in the newspaper, and in news magazines. Here are some examples:

- The average house price in Vancouver compared to that of Montreal.
- The average amount of television watched by college-aged students.
- The average GPA required to be accepted at a college or university in Ontario.

If we consider only the central value in a set of data, or if we compare several sets of data using central values, we may draw an erroneous conclusion. In addition to the central values, we should consider the **dispersion**—often called the *variation* or the *spread*—in the data. As an illustration, suppose the average annual income of marketing executives for Internet-related companies is $80 000, and the average income for executives in pharmaceutical firms is also $80 000. If we looked only at the average incomes, we might wrongly conclude that the two salary distributions are identical or nearly identical. A look at the salary ranges indicates that this conclusion is not correct. The salaries for the marketing executives in the Internet firms range from $70 000 to $90 000, but salaries for the marketing executives in pharmaceuticals range from $40 000 to $120 000. Thus, we conclude that although the average salaries are the same for the two industries, there is much more spread or dispersion in salaries for the pharmaceutical executives. To evaluate the dispersion we will consider the range, the mean deviation, the variance, and the standard deviation.

We begin by discussing measures of location. There is not just one measure of location; in fact, there are many. We will consider five: the arithmetic mean, the weighted mean, the median, the mode, and the geometric mean. The arithmetic mean is the most widely used and widely reported measure of central tendency. We study the mean as both a population parameter and a sample statistic.

Describing Data: Numerical Measures

The Population Mean

Many studies involve all the values in a population. For example, there are 12 sales associates employed at the Reynolds Road outlet of Carpets by Otto. The mean amount of commission they earned last month was $1345. We consider this a population value because we considered *all* the sales associates. Other examples of a population mean would be: the mean closing price for Nortel Networks Corporation stock for the last 5 days is $3.37; the mean annual rate of return for the last 10 years for Berger Funds is 5.21 percent; and the mean number of hours of overtime worked last week by the six welders in the welding department of Butts Welding Inc. is 6.45 hours.

For raw data, that is, data that has not been grouped in a frequency distribution or a stem-and-leaf display, the population mean is the sum of all the values in the population divided by the number of values in the population. To find the population mean, we use the following formula.

$$\text{Population mean} = \frac{\text{Sum of all the values in the population}}{\text{Number of values in the population}}$$

Instead of writing out in words the full directions for computing the population mean (or any other measure), it is more convenient to use the shorthand symbols of mathematics. The mean of a population using mathematical symbols is:

POPULATION MEAN $\quad\quad\quad \mu = \frac{\Sigma X}{N} \quad\quad\quad$ [3–1]

where:

- μ represents the population mean. It is the Greek lowercase letter "mu."
- N is the number of items in the population.
- X represents any particular value.
- Σ is the Greek capital letter "sigma" and indicates the operation of adding.
- ΣX is the sum of the X values.

Any measurable characteristic of a population is called a **parameter**. The mean of a population is a parameter.

PARAMETER A characteristic of a population.

Statistics in Action

The average Canadian woman is 163 cm tall and weighs 65.8 kg. The average Canadian man is 178 cm, and weighs 83.2 kg. The average age at which Canadian women marry for the first time is 28, and for a Canadian man, it is 30. Canadian families need an average of 16.1 weeks to pay for income tax and 9.6 weeks to pay for food. The average Canadian couple will have 1.7 children. The average life expectancy of a Canadian boy in 1951 was 66 years, while that of a girl was 71 years. For those born in 2002, the average life expectancy for both males and females has increased by 11 years.

EXAMPLE

There are 15 teams in the Eastern Conference of the NHL. Listed below is the number of goals scored by each team in the 2003–2004 season.

Team	Goals Scored	Team	Goals Scored
Ottawa Senators	262	New York Rangers	206
Buffalo Sabres	220	New York Islanders	237
Toronto Maple Leafs	242	Washington Capitals	186
Boston Bruins	209	Carolina Hurricanes	172
Montreal Canadiens	208	Florida Panthers	188
New Jersey Devils	213	Atlanta Thrashers	214
Philadelphia Flyers	229	Tampa Bay Lightning	245
Pittsburgh Penguins	190		

Is this a sample or a population? What is the arithmetic mean number of goals scored?

Solution

This is a population if the researcher is considering only the teams in the Eastern Conference; otherwise, it is a sample. We add the number of goals scored for each of the 15 teams. The total number of goals scored for the 15 teams is 3221. To find the arithmetic mean, we divide this total by 15. Therefore, the arithmetic mean is 215, found by 3221/15. Using formula (3–1):

$$\mu = \frac{262 + 220 + \cdots + 245}{15} = \frac{3221}{15} = 215$$

How do we interpret the value of 215? The typical number of goals scored by a team in the Eastern Conference in the 2003–2004 season is 215.

The Sample Mean

As explained in Chapter 1, we often select a sample from a population to find something about a specific characteristic of the population. The quality assurance department, for example, needs to be assured that the ball bearings being produced have an acceptable outside diameter. It would be very expensive and time consuming to check the outside diameter of all the bearings produced. Therefore, a sample of five bearings is selected and the mean outside diameter of the five bearings is calculated to estimate the mean diameter of all the bearings.

For raw data, that is, ungrouped data, *the mean is the sum of all the sampled values divided by the total number of sampled values.* To find the mean for a sample:

Mean of ungrouped sample data

$$\text{Sample mean} = \frac{\text{Sum of all the values in the sample}}{\text{Number of values in the sampe}}$$

The mean of a sample and the mean of a population are computed in the same way, but the shorthand notation used is different. The formula for the mean of a *sample* is:

SAMPLE MEAN $\bar{X} = \frac{\Sigma X}{n}$ [3–2]

where:

$\bar{X}$ is the sample mean. It is read "X bar."
n is the number in the sample.

The mean of a sample, or any other measure based on sample data, is called a **statistic**. If the mean outside diameter of a sample of five ball bearings is 0.625 inches, this is an example of a statistic.

STATISTIC A characteristic of a sample.

EXAMPLE

A sample of average retail prices in 2003 for regular unleaded gasoline at full service filling stations, in cents per litre, is listed below:

Urban Centre	¢/L
St. John's, Newfoundland and Labrador	85.9
Halifax, Nova Scotia	81.1
Saint John, New Brunswick	81.3
Montreal, Quebec	78.8
Toronto, Ontario	72.7
Winnipeg, Manitoba	68.2
Saskatoon, Saskatchewan	75.9
Calgary, Alberta	67.3
Vancouver, British Columbia	79.2

Source: Adapted from Statistics Canada CANSIM database, http://cansim2.statcan.ca, Table 326-009, March 7, 2005.

What is the arithmetic mean gas price for this sample of nine Canadian Urban Centres?

Describing Data: Numerical Measures 57

Solution

Using formula (3–2), the sample mean is:

$$\text{Sample mean} = \frac{\text{Sum of all the values in the sample}}{\text{Number of values in the sample}}$$

$$\bar{X} = \frac{\Sigma X}{n} = \frac{85.9 + 81.1 + \cdots + 79.2}{9} = \frac{690.4}{9} = 76.7$$

The arithmetic mean gas price for the sample of nine Canadian Urban Centres is 76.7 cents.

Statistics in Action

Most colleges report the "average class size." This information can be misleading because average class size can be found several ways. If we find the number of students *in each class* at a particular school, the result is the mean number of students per class. If we compiled a list of the class sizes for each student and find the mean class size, we might find the mean to be quite different. One school found the mean number of students in each of their 747 classes to be 40. But when they found the mean from a list of the class sizes of each student it was 147. Why the disparity? Because there are fewer students in the small classes and a larger number of students in the larger class, which has the effect of increasing the mean class size when it is calculated this way. A school could reduce

(Continued on Next Page)

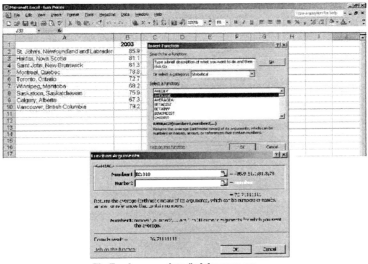

The Excel commands to find the mean are:
1. From the tool bar, select the **Paste Function**, or use **Insert, Function**.
2. From the Function category list, select **Statistical**. In the Function name list, select **AVERAGE**. Click OK. A dialog box opens.
3. Enter the range B2:B10 in the **Number1** box. The answer appears in the dialog box. Click OK.

The Properties of the Arithmetic Mean

The arithmetic mean is a widely used measure of central tendency. It has several important properties:

1. Every set of interval-level or ratio-level data has a mean. (Recall from Chapter 1 that ratio-level data include such data as ages, incomes, and weights, with the distance between numbers being constant.)
2. All the values are included in computing the mean.
3. A set of data has only one mean. The mean is unique. (Later in the chapter we will discover an average that might appear twice, or more than twice, in a set of data.)
4. The mean is a useful measure for comparing two or more sets of data. It can, for example, be used to compare the performance of the production employees on the first shift at the Chrysler transmission plant with the performance of those on the second shift.
5. The arithmetic mean is the only measure of central tendency where *the sum of the deviations of each value from the mean will always be zero.* Expressed symbolically:

$$\Sigma(X - \bar{X}) = 0$$

Chapter 3

(Continued) this mean class size for each student by reducing the number of students in each class. That is, they could cut out the large first-year classes.

Mean as a balance point

As an example, the mean of 3, 8, and 4 is 5. Then:

$$\Sigma(X - \bar{X}) = (3 - 5) + (8 - 5) + (4 - 5)$$
$$= -2 + 3 - 1$$
$$= 0$$

Thus, we can consider the mean as a balance point for a set of data. To illustrate, we have a long board with the numbers 1, 2, 3, ..., n evenly spaced on it. Suppose three bars of equal weight were placed on the board at numbers 3, 4, and 8, and the balance point was set at 5, the mean of the three numbers. We would find that the board balanced perfectly! The deviations below the mean (−3) are equal to the deviations above the mean (+3). Shown schematically:

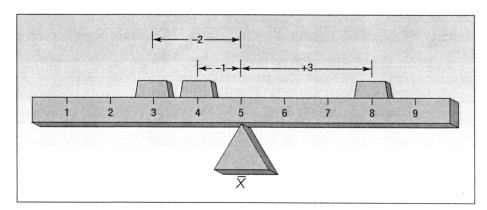

Mean unduly affected by unusually large or small values

The mean does have a major weakness. Recall that the mean uses the value of every item in a sample, or population, in its computation. If one or two of these values are either extremely large or extremely small, the mean might not be an appropriate average to represent the data. For example, suppose the annual incomes of a small group of stockbrokers at Merrill Lynch are $62 900, $61 600, $62 500, $60 800, and $1.2 million. The mean income is $289 560. Obviously, it is not representative of this group, because all but one broker has an income in the $60 000 to $63 000 range. One income ($1.2 million) unduly affects the mean.

Self-Review 3–1

1. The annual incomes of a sample of several middle-management employees at Westinghouse are: $62 900, $69 100, $58 300, and $76 800.
   (a) Find the sample mean.
   (b) Is the mean you computed in (a) a statistic or a parameter? Why?
   (c) What is your best estimate of the population mean?
2. All the students in advanced Computer Science 411 are considered the population. Their course grades are 92, 96, 61, 86, 79, and 84.
   (a) Compute the mean course grade.
   (b) Is the mean you computed in (a) a statistic or a parameter? Why?

Exercises

The answers to the odd-numbered exercises are at the end of the book.

1. Compute the mean of the following population values: 6, 3, 5, 7, 6.
2. Compute the mean of the following population values: 7, 5, 7, 3, 7, 4.
3. a. Compute the mean of the following sample values: 5, 9, 4, 10.
   b. Show that $\Sigma(X - \bar{X}) = 0$.

Describing Data: Numerical Measures 59

4. **a.** Compute the mean of the following sample values: 1.3, 7.0, 3.6, 4.1, 5.0.
   b. Show that $\Sigma(X - \bar{X}) = 0$.
5. Compute the mean of the following sample values: 16.25, 12.91, 14.58.
6. Compute the mean hourly wage paid to carpenters who earned the following wages: $15.40, $20.10, $18.75, $22.76, $30.67, $18.00.

For Exercises 7–10, (a) compute the arithmetic mean and (b) indicate whether it is a statistic or a parameter.

7. There are 10 salespeople employed by Midtown Ford. The numbers of new cars sold last month by the respective salespeople were: 15, 23, 4, 19, 18, 10, 10, 8, 28, 19.
8. The accounting department at a mail-order company counted the following numbers of incoming calls per day to the company's toll-free number during the first 7 days in May 2001: 14, 24, 19, 31, 36, 26, 17.
9. The Cambridge Power and Light Company selected 20 residential customers at random. Following are the amounts, to the nearest dollar, the customers were charged for electrical service last month:

54	48	58	50	25	47	75	46	60	70
67	68	39	35	56	66	33	62	65	67

10. The Human Resources Director at Ford began a study of the overtime hours in the Inspection Department. A sample of 15 workers was selected at random and records showed they worked the following number of overtime hours last month.

13	13	12	15	7	15	5	12
6	7	12	10	9	13	12	

The Weighted Mean

The weighted mean is a special case of the arithmetic mean. It occurs when there are several observations of the same value. To explain, suppose the nearby Wendy's Restaurant sold medium, large, and Biggie-sized soft drinks for $1.19, $1.39, and $1.59, respectively. Of the last 10 drinks sold, 3 were medium, 4 were large, and 3 were Biggie-sized. To find the mean price of the last 10 drinks sold, we could use formula (3–2).

$$\bar{X} = \frac{\$1.19 + \$1.19 + \$1.19 + \$1.39 + \$1.39 + \$1.39 + \$1.39 + \$1.59 + \$1.59 + \$1.59}{10}$$

$$= \frac{\$13.90}{10} = \$1.39$$

The mean selling price of the last 10 drinks is $1.39.

An easier way to find the mean selling price is to determine the weighted mean. That is, we multiply each observation by the number of times it happens. We will refer to the weighted mean as $\bar{X}_w$. This is read "X bar sub w."

$$\bar{X}_w = \frac{3(\$1.19) + 4(\$1.39) + 3(\$1.59)}{10} = \frac{\$13.90}{10} = \$1.39$$

In general the weighted mean of a set of numbers designated $X_1, X_2, X_3, \ldots, X_n$ with the corresponding weights $w_1, w_2, w_3, \ldots, w_n$ is computed by:

WEIGHTED MEAN
$$\bar{X}_w = \frac{w_1 X_1 + w_2 X_2 + w_3 X_3 + \cdots + w_n X_n}{w_1 + w_2 + w_3 + \cdots + w_n} \qquad [3\text{--}3]$$

This may be shortened to:

$$\bar{X}_w = \frac{\Sigma(wX)}{\Sigma w}$$

EXAMPLE

Professor Hunking just marked a quiz for his finance class. The grades follow. What is the mean grade on the quiz?

Grade on Quiz	4	5	6	7	8	9	10
Number of Students	3	3	7	12	9	5	2

Solution

To find the mean grade, we multiply each of the grades by the number of students earning that grade. Using formula (3–3), the mean grade is

$$\bar{X}_w = \frac{3(4) + 3(5) + 7(6) + 12(7) + 9(8) + 5(9) + 2(10)}{3 + 3 + 7 + 12 + 9 + 5 + 2} = \frac{290}{41} = 7.073$$

The weighted mean grade is rounded to 7.1.

Self-Review 3–2

Springers sold 95 Antonelli men's suits for the regular price of $400. For the spring sale the suits were reduced to $200 and 126 were sold. At the final clearance, the price was reduced to $100 and the remaining 79 suits were sold.

(a) What was the weighted mean price of an Antonelli suit?
(b) Springers paid $200 a suit for the 300 suits. Comment on the store's profit per suit if a salesperson receives a $25 commission for each one sold.

Exercises

11. In June an investor purchased 300 shares of Oracle stock (an information technology company) at $20 per share. In August she purchased an additional 400 shares at $25 per share. In November she purchased an additional 400 shares, but the stock declined to $23 per share. What is the weighted mean price per share?
12. The Bookstall, Inc., is a specialty bookstore concentrating on used books sold via the internet. Paperbacks are $1.00 each, and hardcover books are $3.50. Of the 50 books sold last Tuesday morning, 40 were paperback and the rest were hardcover. What was the weighted mean price of a book?
13. The Loris Healthcare System employs 200 persons on the nursing staff. Fifty are nurse's aides, 50 are practical nurses, and 100 are registered nurses. Nurse's aides receive $16 per hour, practical nurses $20 per hour, and registered nurses $28 per hour. What is the weighted mean hourly wage?
14. Andrews and Associates specialize in corporate law. They charge $100 per hour for researching a case, $75 per hour for consultations, and $200 per hour for writing a brief. Last week one of the associates spent 10 hours consulting with her client, 10 hours researching the case, and 20 hours writing the brief. What was the weighted mean hourly charge for her legal services?

The Median

We have stressed that for data containing one or two very large or very small values, the arithmetic mean may not be representative. The centre point for such data can be better described using a measure of location called the **median.**

To illustrate the need for a measure of central tendency other than the arithmetic mean, suppose you are seeking to buy a condominium in St. John's, Newfoundland. Your real estate agent says that the average price of the units currently available is $110 000. Would you still want to look? If you had budgeted your maximum purchase price between $60 000 and $75 000, you might think they are out of your price range. However, checking the individual prices of the units might change your mind. They are $60 000, $65 000, $70 000, $80 000, and a deluxe penthouse costs $275 000. The arithmetic mean price is $110 000, as the real estate agent reported, but one price ($275 000) is pulling the arithmetic mean upward, causing it to be an unrepresentative average. It does seem that a price between $65 000 and $75 000 is a more typical or representative average, and it is. In cases such as this, the median provides a more valid measure of location.

Describing Data: Numerical Measures

> **MEDIAN** The midpoint of the values after they have been ordered from the smallest to the largest, or the largest to the smallest.

The data must be at least ordinal level of measurement. The median price of the units available is $70 000. To determine this, we ordered the prices from low ($60 000) to high ($275 000) and selected the middle value ($70 000).

Prices Ordered from Low to High ($)		Prices Ordered from High to Low ($)
60 000		275 000
65 000		80 000
70 000	← Median →	70 000
80 000		65 000
275 000		60 000

Median unaffected by extreme values

Note that there are the same number of prices below the median of $70 000 as above it. There are as many values below the median as above. The median is, therefore, unaffected by extremely low or high prices. Had the highest price been $90 000, or $300 000, or even $1 million, the median price would still be $70 000. Likewise, had the lowest price been $20 000 or $50 000, the median price would still be $70 000.

In the previous illustration there is an *odd* number of observations (five). How is the median determined for an *even* number of observations? As before, the observations are ordered. Then we calculate the mean of the two middle observations. Note that for an even number of observations, the median may not be one of the given values.

EXAMPLE

The three-year returns of the six best-performing Canadian Equity funds are listed below. What is the median return? Note: funds with less than a three-year track record and less than $10 million in total assets are excluded.

Name of Fund	*Total Three-year Return (%)
Acuity All Cap 30 Canadian Equity	24.4
Acuity Pooled Canadian Equity	22.6
Northwest Specialty Equity	32.6
R Small Cap Canadian Equity	20.8
Sceptre Equity Growth	33.0
Sprott Canadian Equity	28.9

Source: Morningstar Canada
*Performance data is as of February 28, 2005.

Solution

Note that the number of returns is *even* (6). As before, the returns are first ordered from low to high. Then the two middle returns are identified. The arithmetic mean of the two middle observations gives us the median return. Arranging from low to high:

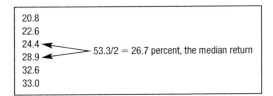

Notice that the median is not one of the values. Also, half of the returns are below the median and half are above it.

Chapter 3

The major properties of the median are:

1. The median is unique; that is, like the mean, there is only one median for a set of data.
2. It is not affected by extremely large or small values and is therefore a valuable measure of location when such values do occur.
3. It can be computed for ratio-level, interval-level, and ordinal-level data. To use a simple illustration, suppose five people rated a new fudge bar. One person thought it was excellent, one rated it very good, one called it good, one rated it fair, and one considered it poor. The median response is "good." Half of the responses are above "good"; the other half are below it.

Median can be determined for all levels of data except nominal

The Excel commands to find the median are:

1. From the tool bar, select the Paste Function, or use Insert, Function.
2. From the Function category list, select Statistical. In the Function name list, select MEDIAN. Click OK. A dialogue box opens.
3. Enter the range in the Number1 box. The answer appears in the dialogue box. Click OK.

The Mode

The **mode** is another measure of central tendency.

> **MODE** The value of the observation that appears most frequently.

The mode is especially useful in describing nominal and ordinal levels of measurement. As an example of its use for nominal-level data, a company has developed five bath oils. Chart 3–1 shows the results of a marketing survey designed to find which bath oil consumers prefer. The largest number of respondents favoured Lamoure, as evidenced by the highest bar. Thus, Lamoure is the mode.

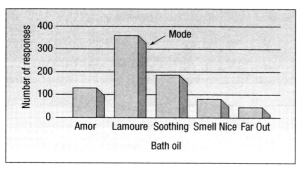

CHART 3–1 Number of Respondents Favouring Various Bath Oils.

EXAMPLE

Average earnings of Canadians 15 years and older with university degrees in selected cities are shown below. (Statistics Canada, Census of Population, 2001)

City	Salary ($)	City	Salary ($)
Calgary	58 000	Regina	46 000
Edmonton	45 000	Saint John	36 000
Halifax	43 000	Greater Sudbury	50 000
Hamilton	52 000	Toronto	56 000
London	48 000	Winnipeg	43 000
Montreal	47 000	Vancouver	46 000
Ottawa-Gatineau	55 000	Victoria	42 000

Source: Adapted from the Statistics Canada website http://www.statcan.ca/english/Pgdb/labor50a.htm; March 29, 2005.

Describing Data: Numerical Measures 63

Solution | A perusal of the earnings reveals that $43 000 and $46 000 appear more often (twice each) than any other amount. Therefore, the modes are $43 000 and $46 000.

The Excel commands to find the mode are similar to those of the mean and median.
1. Select the Paste Function.
2. In the Function name list, select MODE. Click OK. A dialogue box opens.
3. Enter the range in the Number1 box. The answer appears in the dialogue box. Click OK.

In summary, we can determine the mode for all levels of data—nominal, ordinal, interval, and ratio. The mode also has the advantage of not being affected by extremely high or low values.

Disadvantages of the mode

The mode does have a number of disadvantages, however, that cause it to be used less frequently than the mean or median. For many sets of data, there is no mode because no value appears more than once. For example, there is no mode for this set of price data: $19, $21, $23, $20, and $18. Since every value is different, however, it could be argued that every value is the mode. Conversely, for some data sets there is more than one mode. Suppose the ages of a group are 22, 26, 27, 27, 31, 35, and 35. Both the ages 27 and 35 are modes. Thus, this grouping of ages is referred to as *bimodal* (having two modes). One would question the use of two modes to represent the central tendency of this set of age data.

Self-Review 3–3

1. According to Statistics Canada, the average weekly employment insurance benefits for 2004, by category, are: $297, $293, $270, $309, $381, $91, and $366.
   Source: (Adapted from Statistics Canada CANSIM database, http://cansim2.statcan.ca. Table 276–0016; Apr 4, 2005.)
   (a) What is the median monthly benefit?
   (b) How many observations are below the median? Above it?
2. The numbers of work stoppages in the automobile industry for selected months are 6, 0, 10, 14, 8, and 0.
   (a) What is the median number of stoppages?
   (b) How many observations are below the median? Above it?
   (c) What is the modal number of work stoppages?

Exercises

15. What would you report as the modal value for a set of observations if there were a total of:
   a. 10 observations and no two values were the same?
   b. 6 observations and they were all the same?
   c. 6 observations and the values were 1, 2, 3, 3, 4, and 4?

For Exercises 16–18, (a) determine the median and (b) the mode.

16. The following is the number of oil changes for the last 7 days at the Jiffy Lube located at the corner of Elm Street and Fortson Ave.

| 41 | 15 | 39 | 54 | 31 | 15 | 33 |

17. The following is the percent change in net income from 2004 to 2005 for a sample of 12 construction companies in Benton.

| 5 | 1 | −10 | −6 | 5 | 12 | 7 | 8 | 2 | 5 | −1 | 11 |

18. The following are the ages of the 10 people in the video arcade at the Southwyck Shopping Mall at 10 A.M. this morning.

| 12 | 8 | 17 | 6 | 11 | 14 | 8 | 17 | 10 | 8 |

19. Listed below is the average earnings ratio by sex for full-year, full-time workers from 1993 to 2002. (Statistics Canada, CANSIM table 202-0102).

Chapter 3

Year	Women	Men	Ratio (%)	Year	Women	Men	Ratio (%)
1993	33 300	46 200	72.1	1998	35 500	49 200	72.1
1994	33 100	47 500	69.7	1999	34 300	49 400	69.4
1995	33 900	46 400	73.0	2000	35 400	49 400	71.7
1996	33 300	45 800	72.8	2001	35 900	50 500	71.0
1997	33 200	48 000	69.2	2002	36 000	50 500	71.3

   a. What is the median earnings ratio?
   b. What is the modal earnings ratio?
20. Listed below are the total automobile sales (in millions) for the last 14 years. During this period, what was the median number of automobiles sold? What is the mode?

| 9.0 | 8.5 | 8.0 | 9.1 | 10.3 | 11.0 | 11.5 | 10.3 | 10.5 | 9.8 | 9.3 | 8.2 | 8.2 | 8.5 |

Computer Solution

We can use Excel or MegaStat to find the measures of central tendency at once.

EXAMPLE

Table 2–1 shows the list prices of 90 homes listed in Regina and surrounding area. Determine the mean, median and modal list price.

Solution

The mean, median and modal list prices are reported in the following Excel output. There are 90 list prices, so calculations with a calculator would be tedious and prone to error.

List Price	
Mean	192712.2222
Standard Error	9158.045906
Median	187400
Mode	139900
Standard Deviation	86880.85194
Sample Variance	7548282433
Kurtosis	0.002855765
Skewness	0.773358677
Range	356400
Minimum	62500
Maximum	418900
Sum	17344100
Count	90

The mean list price is $192 712, the median is $187 400 and the mode is $139 900. The mean is somewhat distorted by several listings that are over $300 000 of which 2 are over $400 000. The mode is much lower than both the mean and median, and so the median value is most representative of the data. We can then conclude that the typical home was listed at about $187 400. We can also see from the output that the count is 90, indicating that the sample consisted of 90 list prices.

The Excel commands to create the descriptive statistics for the list prices follow.

1. Open Excel and the Excel file Table 02-1 from the DataSets on the CD provided.
2. From the menu bar, select Tools, Data Analysis, and Descriptive Statistics; then click OK.

Describing Data: Numerical Measures 65

3. Enter A1:A91 as the Input Range. For Grouped By, select Columns to indicate that your data is in a column; select Labels in First Row to indicate that you have the label List Price in the first cell of the input range. Place the output in the same worksheet by entering C3 in the Output Range.
4. Select the Summary statistics box; click OK.

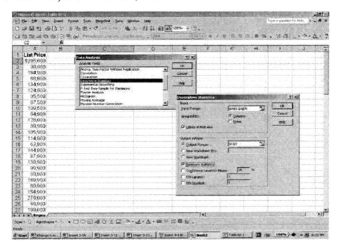

The MegaStat commands to create the descriptive statistics for the list prices follow.

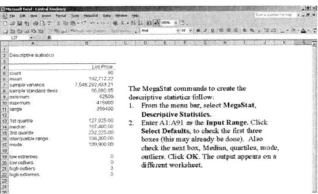

The MegaStat commands to create the descriptive statistics follow:
1. From the menu bar, select **MegaStat, Descriptive Statistics.**
2. Enter A1:A91 as the **Input Range.** Click **Select Defaults,** to check the first three boxes (this may already be done). Also check the next box, Median, quartiles, mode, outliers. Click OK. The output appears on a different worksheet.

The Relative Positions of the Mean, Median, and Mode

For a symmetric, mound-shaped distribution, mean, median, and mode are equal.

Refer to the frequency polygon in Chart 3–2. It is a symmetric distribution, which is also mound-shaped. This distribution *has the same shape on either side of the centre*. If the polygon were folded in half, the two halves would be identical. For this symmetric distribution, the mode, median, and mean are located at the centre and are always equal. They are all equal to 20 years in Chart 3–2. We should point out that there are symmetric distributions that are not mound-shaped.

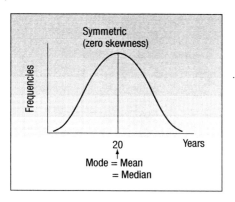

CHART 3–2 A Symmetric Distribution

A skewed distribution is not symmetrical.

The number of years corresponding to the highest point of the curve is the *mode* (20 years). Because the frequency curve is symmetrical, the *median* corresponds to the point where the distribution is cut in half (20 years). The total number of frequencies representing many years is offset by the total number representing few years, resulting in an *arithmetic mean* of 20 years. Logically, any of the three measures would be appropriate to represent this distribution.

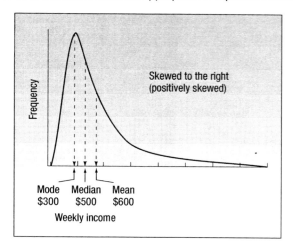

CHART 3–3 A Positively Skewed Distribution

If a distribution is nonsymmetrical, or **skewed,** the relationship among the three measures changes. In a **positively skewed distribution,** the arithmetic mean is the largest of the three measures. Why? Because the mean is influenced more than the median or mode by a few extremely high values. The median is generally the next largest measure in a positively skewed frequency distribution. The mode is the smallest of the three measures.

Describing Data: Numerical Measures

If the distribution is highly skewed, such as the weekly incomes in Chart 3–3, the mean would not be a good measure to use. The median and mode would be more representative.

Conversely, in a distribution that is **negatively skewed,** the mean is the lowest of the three measures. The mean is, of course, influenced by a few extremely low observations. The median is greater than the arithmetic mean, and the modal value is the largest of the three measures. Again, if the distribution is highly skewed, such as the distribution of tensile strengths shown in Chart 3–4, the mean should not be used to represent the data.

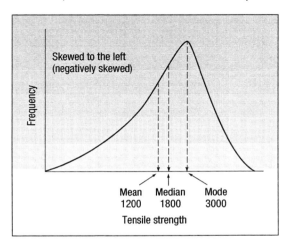

CHART 3–4 A Negatively Skewed Distribution

Self-Review 3–4

The weekly sales from a sample of Hi-Tec electronic supply stores were organized into a frequency distribution. The mean of weekly sales was computed to be $105 900, the median $105 000, and the mode $104 500.

(a) Sketch the sales in the form of a smoothed frequency polygon. Note the location of the mean, median, and mode on the X-axis.
(b) Is the distribution symmetrical, positively skewed, or negatively skewed? Explain.

The Geometric Mean

The geometric mean is never greater than the arithmetic mean.

The geometric mean is useful in finding the average of percentages, ratios, indexes, or growth rates. It has a wide application in business and economics because we are often interested in finding the percentage changes in sales, salaries, or economic figures, such as the Gross Domestic Product, which compound or build on each other. The geometric mean of a set of n positive numbers is defined as the nth root of the product of the n values. The formula for the geometric mean is written:

| **GEOMETRIC MEAN** | $GM = \sqrt[n]{(X_1)(X_2)\cdots(X_n)}$ | [3–4] |

The geometric mean will always be less than or equal to (never more than) the arithmetic mean. Also all the data values must be positive.

As an example of the geometric mean, suppose you receive a 5 percent increase in salary this year and a 15 percent increase next year. The average percent increase is 9.886, not 10.0. Why is this so? We begin by calculating the geometric mean. Recall, for example, that a 5 percent increase in salary is 105 percent or 1.05. We will write it as 1.05.

$$GM = \sqrt{(1.05)(1.15)} = 1.09886$$

Chapter 3

This can be verified by assuming that your monthly earning was $3000 to start and you received two increases of 5 percent and 15 percent.

$$\text{Raise 1} = \$3000 (.05) = \$150.00$$
$$\text{Raise 2} = \$3150 (.15) = \underline{472.50}$$
$$\text{Total} \qquad \$622.50$$

Your total salary increase is $622.50. This is equivalent to:

$$\$3000.00 (.09886) = \$296.58$$
$$\$3296.58 (.09886) = \underline{325.90}$$
$$\$622.48 \text{ is about } \$622.50$$

The following example shows the geometric mean of several percentages.

EXAMPLE

The return on investment earned by Atkins Construction Company for four successive years was: 30 percent, 20 percent, −40 percent, and 200 percent. What is the geometric mean rate of return on investment?

Solution

The number 1.3 represents the 30 percent return on investment, which is the "original" investment of 1.0 plus the "return" of 0.3. The number 0.6 represents the loss of 40 percent, which is the original investment of 1.0 reduced by 40 percent (−0.4). This calculation assumes the total return each period is reinvested or becomes the base for the next period. In other words, the base for the second period is 1.3 and the base for the third period is (1.3)(1.2) and so forth.

Then the geometric mean rate of return is 29.4 percent, found by

$$GM = \sqrt[n]{(X_1)(X_2)\cdots(X_n)} = \sqrt[4]{(1.3)(1.2)(0.6)(3.0)} = 1.294$$

The geometric mean is the fourth root of 2.808. So, the average rate of return (compound annual growth rate) is 29.4 percent. In other words, if Dunking Construction started with the same capital that Atkins had and earned a return on investment of 29.4 percent per year for four successive years, they would be in exactly the same position.

Notice also that if you compute the arithmetic mean [(30 + 20 − 40 + 200)/4 = 52.5], you would have a much larger number, which would overstate the true rate of return!

A second application of the geometric mean is to find an average percent increase over a period of time. For example, if you earned $30 000 in 1992 and $50 000 in 2002, what is your annual rate of increase over the period? The rate of increase is determined from the following formula.

AVERAGE PERCENT INCREASE OVER TIME
$$GM = \sqrt[n]{\frac{\text{Value at end of period}}{\text{Value at beginning of period}}} - 1 \qquad [3\text{--}5]$$

In the above box n is the number of periods. An example will show the details of finding the average annual percent increase.

EXAMPLE

The population of Alberta grew from 2 667 448 in 1993 to 3 153 723 in 2003. What was the average annual rate of percentage increase during the period?

Solution

There are 10 years between 1993 and 2003 so $n = 10$. The formula (3–5) for the geometric mean as applied to this type of problem is:

$$GM = \sqrt[n]{\frac{\text{Value at end of period}}{\text{Value at beginning of period}}} - 1$$

$$= \sqrt[10]{\frac{3\,153\,723}{2\,667\,448}} - 1 = 1.0169 - 1 = 0.0169$$

The final value is 0.0169. So the annual rate of increase is 1.69 percent. This means that the rate of population growth in the province is 1.69 percent per year.

Self-Review 3–5

1. The annual dividends, in percent, for the last 4 years at Combs Cosmetics are: 4.91, 5.75, 8.12, and 21.60.
   (a) Find the geometric mean dividend.
   (b) Find the arithmetic mean dividend.
   (c) Is the arithmetic mean equal to or greater than the geometric mean?
2. Production of Cablos trucks increased from 23 000 units in 1985 to 120 520 units in 2005. Find the geometric mean annual percent increase.

Exercises

21. Compute the geometric mean of the following percent increases: 8, 12, 14, 26, and 5.
22. Compute the geometric mean of the following percent increases: 2, 8, 6, 4, 10, 6, 8, and 4.
23. Listed below is the percent increase in sales for the MG Corporation over the last 5 years. Determine the geometric mean percent increase in sales over the period.

9.4	13.8	11.7	11.9	14.7

24. In 1978, the CPI for Canada was 43.6, but in 1988, the CPI was 84.8. What was the geometric mean annual increase for the period?
25. The population of Canada has increased from 21 961 999 in 1971 to 31 629 677 in 2003. What was the geometric mean annual increase for the period?
26. Gas prices in St. John's, Newfoundland and Labrador have increased from 64.1 cents per litre for regular unleaded in 1990, to 85.9 cents per litre in 2003. What was the geometric mean annual increase for the period?
27. In 1999 there were 42.0 million pager subscribers. By 2004 the number of subscribers increased to 70.0 million. What is the geometric mean annual increase for the period?
28. Tuition fees for dentistry in Canada increased from $7863 for the 1999–2000 academic year to $11 733 in the 2003–2004 academic year. What was the geometric mean annual increase for the period?

Why Study Dispersion?

A measure of location, such as the mean or the median, only describes the centre of the data. It is valuable from that standpoint, but it does not tell us anything about the spread of the data. For example, if your nature guide told you that the river ahead averaged 1 m in depth, would you cross it without additional information? Probably not. You would want to know something about the variation in the depth. Is the maximum depth of the river 1.25 m and the minimum 0.5 m? If that is the case, you would probably agree to cross. What if you learned the river depth ranged from 0.50 m to 2 m? Your decision would probably be not to cross. Before making a decision about crossing the river, you want information on both the typical depth and the dispersion in the depth of the river.

A small value for a measure of dispersion indicates that the data are clustered closely, say, around the arithmetic mean. The mean is therefore considered representative of the data. Conversely, a large measure of dispersion indicates that the mean is not reliable. Refer to Chart 3–5. The 100 employees of Hammond Iron Works, Inc., a steel fabricating company, are organized into a histogram based on the number of years of employment with the company. The mean is 4.9 years, but the spread of the data is from 6 months to 16.8 years. The mean of 4.9 years is not very representative of all the employees.

The average is not representative because of the large spread.

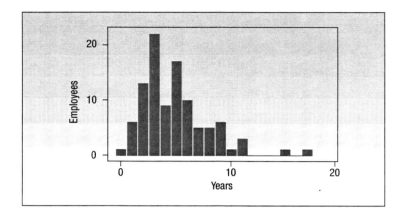

CHART 3–5 Histogram of Years of Employment at Hammond Iron Works, Inc.

A second reason for studying the dispersion in a set of data is to compare the spread in two or more distributions. Suppose, for example, that the new PDM/3 computer is assembled in Kanata and also in Waterloo. The arithmetic mean hourly output in both the Kanata plant and the Waterloo plant is 50. Based on the two means, one might conclude that the distributions of the hourly outputs are identical. Production records for 9 hours at the two plants, however, reveal that this conclusion is not correct (see Chart 3–6). Kanata production varies from 48 to 52 assemblies per hour. Production at the Waterloo plant is more erratic, ranging from 40 to 60 per hour. Therefore, the hourly output for Kanata is clustered near the mean of 50; the hourly output for Waterloo is more dispersed.

A measure of dispersion can be used to evaluate the reliability of two or more measures of location.

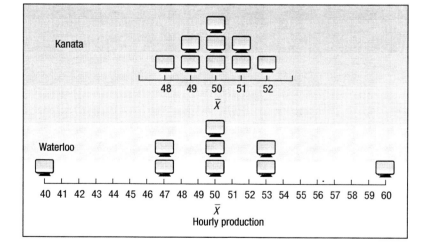

CHART 3–6 Hourly Production of Computers at the Kanata and Waterloo Plants

Measures of Dispersion

We will consider several measures of dispersion. The range is based on the largest and the smallest values in the data set. The mean deviation, the variance, and the standard deviation are all based on deviations from the arithmetic mean.

Range

The simplest measure of dispersion is the **range.** It is the difference between the largest and the smallest values in a data set. In the form of an equation:

RANGE Range = Largest value − Smallest value [3–6]

The range is widely used in statistical process control (SPC) applications because it is very easy to calculate and understand.

EXAMPLE

Refer to Chart 3–6. Find the range in the number of computers produced per hour for the Kanata and the Waterloo plants. Interpret the two ranges.

Solution

The range of the hourly production of computers at the Kanata plant is 4, found by the difference between the largest hourly production of 52 and the smallest of 48. The range in the hourly production for the Waterloo plant is 20 computers, found by 60 − 40. We therefore conclude that (1) there is less dispersion in the hourly production in the Kanata plant than in the Waterloo plant because the range of 4 computers is less than a range of 20 computers, and (2) the production is clustered more closely around the mean of 50 at the Kanata plant than at the Waterloo plant (because a range of 4 is less than a range of 20). Thus, the mean production in the Kanata plant (50 computers) is a more representative measure of location than the mean of 50 computers for the Waterloo plant.

Mean Deviation

A serious defect of the range is that it is based on only two values, the highest and the lowest; it does not take into consideration all of the values. The **mean deviation** does. It measures the mean amount by which the values in a population, or sample, vary from their mean. In terms of a definition:

MEAN DEVIATION The arithmetic mean of the absolute values of the deviations from the arithmetic mean.

In terms of a formula, the mean deviation, designated MD, is computed for a sample by:

MEAN DEVIATION $$MD = \frac{\Sigma |X - \bar{X}|}{n}$$ [3–7]

where:

X is the value of each observation.
$\bar{X}$ is the arithmetic mean of the values.
n is the number of observations in the sample.
$\|\,\|$ indicates the absolute value.

Why do we ignore the signs of the deviations from the mean? If we didn't, the positive and negative deviations from the mean would exactly offset each other, and the mean deviation would always be zero. Such a measure (zero) would be a useless statistic.

EXAMPLE

The numbers of cappuccinos sold per hour at the Starbucks location in the Orange County Airport between 4 P.M.–9 P.M. were: 103, 97, 101, 106, and 103. Determine the mean deviation and interpret.

Chapter 3

Solution

The mean deviation is the mean of the amounts (ignoring their signs) that individual observations differ from the arithmetic mean. To find the mean deviation of a set of data, we begin by finding the arithmetic mean. The mean number of cappuccinos is 102, found by (103 + 97 + 101 + 106 + 103)/5. Next we find the amount by which each observation differs from the mean. Then we sum these differences, ignoring the signs, and divide the sum by the number of observations. The result is the mean amount the observations differ from the mean. A small value for the mean deviation indicates the mean is representative of the data, whereas a large value for the mean deviation indicates a large dispersion in the data. Below are the details of the calculations using formula (3–7).

Number of Cappuccinos Sold	$(X - \bar{X})$	Absolute Deviation
103	(103 − 102) = 1	1
97	(97 − 102) = −5	5
101	(101 − 102) = −1	1
106	(106 − 102) = 4	4
103	(103 − 102) = 1	1
	Total	12

$$MD = \frac{\Sigma |X - \bar{X}|}{n} = \frac{12}{5} = 2.4$$

The mean deviation is 2.4 cappuccinos per day. The number of cappuccinos deviates, on average, by 2.4 cappuccinos from the mean of 102 cappuccinos per day.

Advantages of mean deviation

The mean deviation has two advantages. First, it uses all the values in the computation. Recall that the range uses only the highest and the lowest values. Second, it is easy to understand—it is the average amount by which values deviate from the mean. However, its major drawback is the use of absolute values. Generally, absolute values are difficult to work with, so the mean deviation is not used as frequently as other measures of dispersion, such as the standard deviation.

Self-Review 3–6

The masses of a group of crates being shipped to Ireland are (in kilograms):

| 95 | 103 | 105 | 110 | 104 | 105 | 112 | 90 |

(a) What is the range of the masses?
(b) Compute the arithmetic mean mass.
(c) Compute the mean deviation of the masses.

Exercises

For Exercises 29–34, calculate the (a) range, (b) arithmetic mean, and (c) mean deviation, and (d) interpret the range and the mean deviation.

29. There were five customer service representatives on duty at the Electronic Super Store during last Friday's sale. The numbers of VCRs these representatives sold are: 5, 8, 4, 10, and 3.
30. The Department of Statistics at a local college offers eight sections of basic statistics. Following are the numbers of students enrolled in these sections: 34, 46, 52, 29, 41, 38, 36, and 28.
31. Dave's Automatic Door installs automatic garage door openers. The following list indicates the number of minutes needed to install a sample of 10 doors: 28, 32, 24, 46, 44, 40, 54, 38, 32, and 42.
32. A sample of eight companies in the aerospace industry was surveyed as to their return on investment last year. The results are (in percent): 10.6, 12.6, 14.8, 18.2, 12.0, 14.8, 12.2, and 15.6.
33. Ten experts rated the taste of a newly developed sushi pizza topped with tuna, rice, and kelp on a scale of 1 to 50. The ratings were: 34, 35, 41, 28, 26, 29, 32, 36, 38, and 40.
34. A sample of the personnel files of eight employees at Acme Carpet Cleaners, Inc. revealed that, during a six-month period, they lost the following numbers of days due to illness: 2, 0, 6, 3, 10, 4, 1, and 2.

Describing Data: Numerical Measures

Variance and standard deviation are based on squared deviations from the mean.

Variance and Standard Deviation

The **variance** and **standard deviation** are also based on deviations from the mean. However, instead of using the absolute value of the deviations, the variance, and the standard deviation square the deviations.

VARIANCE The arithmetic mean of the squared deviations from the mean.

Note that the variance is nonnegative, and it is zero only if all observations are the same.

STANDARD DEVIATION The square root of the variance.

Population Variance The formulas for the population variance and the sample variance are slightly different. The population variance is considered first. (Recall that a population is the totality of all observations being studied.) The **population variance** is found by:

POPULATION VARIANCE $$\sigma^2 = \frac{\Sigma(X - \mu)^2}{N}$$ [3–8]

where:

σ^2 is the symbol for the population variance (σ is the lowercase Greek letter sigma). It is usually referred to as "sigma squared."
X is the value of an observation in the population.
μ is the arithmetic mean of the population.
N is the number of observations in the population.

Note the steps in computing the population variance:

1. Begin by finding the mean.
2. Next, find the difference between each observation and the mean.
3. Square the difference.
4. Sum all of the squared differences.
5. Divide the sum by the total number of observations.

For populations whose values are near the mean, the variance will be small. For populations whose values are dispersed from the mean, the population variance will be large.

The variance overcomes the problem of the range by using all the values in the population, whereas the range uses only the largest and smallest. We overcome the issue of absolute values in the mean deviation by squaring the differences. Squaring the differences between each observation and the mean will always result in non-negative values, so the variance will never be negative and will be zero only when all values in the data set are the same.

EXAMPLE

The ages of all the patients in the isolation ward of Mountainview Hospital are 38, 26, 13, 41, and 22 years. What is the population variance?

Solution

Age (X)	$X - \mu$	$(X - \mu)^2$
38	+10	100
26	−2	4
13	−15	225
41	+13	169
22	−6	36
140	0*	534

$$\mu = \frac{\Sigma X}{N} = \frac{140}{5} = 28$$

$$\sigma^2 = \frac{\Sigma(X - \mu)^2}{N} = \frac{534}{5} = 106.8$$

*Sum of the deviations from mean must equal zero.

Chapter 3

Like the range and the mean deviation, the variance can be used to compare the dispersion in two or more sets of observations. For example, the variance for the ages of the patients in isolation was just computed to be 106.8. If the variance in the ages of the cancer patients in the hospital is 342.9, we conclude that (1) there is less dispersion in the distribution of the ages of patients in isolation than in the age distribution of all cancer patients (because 106.8 is less than 342.9); and (2) the ages of the patients in isolation are clustered more closely about the mean of 28 years than the ages of those in the cancer ward. Thus, the mean age for the patients in isolation is a more representative measure of location than the mean for all cancer patients.

Variance is difficult to interpret because the units are squared.

Population Standard Deviation Both the range and the mean deviation are easy to interpret. The range is the difference between the high and low values of a set of data, and the mean deviation is the mean of the deviations from the mean. However, the variance is difficult to interpret for a single set of observations. The variance of 106.8 for the ages of the patients in isolation is not in terms of years, but rather "years squared."

Standard deviation is in the same units as the data.

There is a way out of this dilemma. By taking the square root of the population variance, we can transform it to the same unit of measurement used for the original data. The square root of 106.8 years-squared is 10.3 years. The square root of the population variance is called the **population standard deviation.**

POPULATION STANDARD DEVIATION
$$\sigma = \sqrt{\frac{\Sigma(X-\mu)^2}{N}}$$ [3–9]

Self-Review 3–7

An office of Price Waterhouse Coopers LLP hired five accounting trainees this year. Their monthly starting salaries were: $3536; $3173; $3448; $3121; and $3622.

(a) Compute the population mean.
(b) Compute the population variance.
(c) Compute the population standard deviation.
(d) Another office hired six trainees. Their mean monthly salary was $3550, and the standard deviation was $250. Compare the two groups.

Exercises

35. Consider these five values a population: 8, 3, 7, 3, and 4.
    a. Determine the mean of the population.
    b. Determine the variance.
36. Consider these six values a population: 13, 3, 8, 10, 8, and 6.
    a. Determine the mean of the population.
    b. Determine the variance.
37. The annual report of Dennis Industries cited these primary earnings per common share for the past 5 years: $2.68, $1.03, $2.26, $4.30, and $3.58. If we assume these are population values, what is:
    a. The arithmetic mean primary earnings per share of common stock?
    b. The variance?
38. Referring to Exercise 37, the annual report of Dennis Industries also gave these returns on stockholder equity for the same five-year period (in percent): 13.2, 5.0, 10.2, 17.5, and 12.9.
    a. What is the arithmetic mean return?
    b. What is the variance?
39. Plywood, Inc. reported these returns on stockholder equity for the past 5 years: 4.3, 4.9, 7.2, 6.7, and 11.6. Consider these as population values.
    a. Compute the range, the arithmetic mean, the variance, and the standard deviation.
    b. Compare the return on stockholder equity for Plywood, Inc. with that for Dennis Industries cited in Exercise 38.
40. The annual incomes of the five vice presidents of TMV Industries are: $125 000; $128 000; $122 000; $133 000; and $140 000. Consider this a population.
    a. What is the range?
    b. What is the arithmetic mean income?

Describing Data: Numerical Measures

c. What is the population variance? The standard deviation?
d. The annual incomes of officers of another firm similar to TMV Industries were also studied. The mean was $129 000 and the standard deviation $8612. Compare the means and dispersions in the two firms.

Sample Variance The formula for the population mean is $\mu = \Sigma X/N$. We just changed the symbols for the sample mean, that is $\bar{X} = \Sigma X/n$. Unfortunately, the conversion from the population variance to the sample variance is not as direct. It requires a change in the denominator. Instead of substituting n (number in the sample) for N (number in the population), the denominator is $n - 1$. Thus the formula for the **sample variance** is:

SAMPLE VARIANCE, DEVIATION FORMULA $$s^2 = \frac{\Sigma(X - \bar{X})^2}{n-1}$$ [3–10]

where:

s^2 is the sample variance.
X is the value of each observation in the sample.
$\bar{X}$ is the mean of the sample.
n is the number of observations in the sample.

Why is this change made in the denominator? Although the use of n is logical, it tends to underestimate the population variance, σ^2. The use of $(n - 1)$ in the denominator provides the appropriate correction for this tendency. For example, in a sample, if we did not have to estimate the population mean using the sample mean, then we could divide by n. But the first estimate we make is the sample mean, and so, we lose a degree of freedom, and hence use $(n - 1)$.

An easier way to compute the variance follows. This formula is much easier to use, even with a hand calculator, because it avoids all but one subtraction. Hence, we recommend formula (3–11) for calculating a sample variance.

SAMPLE VARIANCE, DIRECT FORMULA $$s^2 = \frac{\Sigma X^2 - \frac{(\Sigma X)^2}{n}}{n-1}$$ [3–11]

Statistics in Action

Curtis Joseph of the Detroit Red Wings played in nine games of the 2003–2004 post season of the National Hockey League. He played a total of 518 minutes and was charged with 12 goals against his team making his 60-minute goals against average 1.39 (12/(518/60)). The second ranked player was Evgeni Nabokov of the San Jose Sharks with a 60-minute average of 1.71 goals.

EXAMPLE

The hourly wages for a sample of part-time employees at Fruit Packers, Inc. are: $12, $20, $16, $18, and $19. What is the sample variance?

Solution

The sample variance is computed using two methods. On the left is the deviation method, using formula (3–10). On the right is the direct method, using formula (3–11).

$$\bar{X} = \frac{\Sigma X}{n} = \frac{\$85}{5} = \$17$$

Using squared deviations from the mean:

Hourly Wage ($) X	$ X − $\bar{X}$	^{2} $(X − \bar{X})^2$
12	−5	25
20	3	9
16	−1	1
18	1	1
19	2	4
85	0	40

Using the direct formula:

Hourly Wage ($) X	^{2} X^2
12	144
20	400
16	256
18	324
19	361
85	1485

Chapter 3

$$s^2 = \frac{\Sigma(X - \bar{X})^2}{n-1} = \frac{\$^2 40}{5-1}$$
$$= 10\,\$^2 \text{ (dollars squared)}$$

$$s^2 = \frac{\Sigma X^2 - \frac{(\Sigma X)^2}{n}}{n-1}$$
$$= \frac{1485 - \frac{(85)^2}{5}}{5-1} = \frac{40}{5-1}$$
$$= 10\,\$^2 \text{ (dollars squared)}$$

Sample Standard Deviation The sample standard deviation is used as an estimator of the population standard deviation. As noted previously, the population standard deviation is the square root of the population variance. Likewise, the *sample standard deviation is the square root of the sample variance.* The sample standard deviation is most easily determined by:

STANDARD DEVIATION, DIRECT FORMULA
$$s = \sqrt{\frac{\Sigma X^2 - \frac{(\Sigma X)^2}{n}}{n-1}} \qquad [3\text{-}12]$$

EXAMPLE

The sample variance in the previous example involving hourly wages was computed to be $10\,\$^2$. What is the sample standard deviation?

Solution

The sample standard deviation is $3.16, found by $\sqrt{10}$. Note again that the sample variance is in terms of dollars squared, but taking the square root of 10 gives us $3.16, which is in the same units (dollars) as the original data.

Self-Review 3–8

The masses of the contents of several small aspirin bottles are (in grams): 4, 2, 5, 4, 5, 2, and 6. What is the sample variance? Compute the sample standard deviation.

Exercises

For Exercises 41–45, do the following:

a. Compute the variance using the deviation formula.
b. Compute the variance using the direct formula.
c. Determine the sample standard deviation.

Statistics in Action

An average is a value used to represent all the data. However, often an average does not give the full picture of the set of data. Stockbrokers are often faced with this problem when they are considering two investments, where the

41. Consider these values a sample: 7, 2, 6, 2, and 3.
42. The following five values are a sample: 11, 6, 10, 6, and 7.
43. Dave's Automatic Door, referred to in Exercise 31, installs automatic garage door openers. Based on a sample, following are the times, in minutes, required to install 10 doors: 28, 32, 24, 46, 44, 40, 54, 38, 32, and 42.
44. The sample of eight companies in the aerospace industry, referred to in Exercise 32, was surveyed as to their return on investment last year. The results are: 10.6, 12.6, 14.8, 18.2, 12.0, 14.8, 12.2, and 15.6.
45. Trout, Inc. feeds fingerling trout in special ponds and markets them when they attain a certain weight. A sample of 10 trout were isolated in a pond and fed a special food mixture, designated RT-10. At the end of the experimental period, the masses of the trout were (in grams): 124, 125, 125, 123, 120, 124, 127, 125, 126, and 121.

Describing Data: Numerical Measures

mean rate of return is the same. They usually calculate the standard deviation of the rates of return to assess the risk associated with the two investments. The investment with the larger standard deviation is considered to have the greater risk. In this context the standard deviation plays a vital part in making critical decisions regarding the composition of an investor's portfolio.

46. Refer to Exercise 45. Another special mixture, AB-4, was used in another pond. The mean of a sample was computed to be 126.9 g, and the standard deviation 1.2 g. Which food results in a more uniform mass?

Interpretation and Uses of the Standard Deviation

The standard deviation is commonly used as a measure to compare the spread in two or more sets of observations. For example, the price of two stocks may have about the same mean, but different standard deviations. The stock with the larger standard deviation has more variability in its mean price, and therefore, could be considered more risky.

Chebyshev's Theorem

We have stressed that a small standard deviation for a set of values indicates that these values are located close to the mean. Conversely, a large standard deviation reveals that the observations are widely scattered about the mean. The Russian mathematician P. L. Chebyshev (1821–1894) developed a theorem that allows us to determine the minimum proportion of the values that lie within a specified number of standard deviations of the mean. For example, according to Chebyshev's theorem, at least three of four values, or 75 percent, must lie between the mean plus two standard deviations and the mean minus two standard deviations. This relationship applies regardless of the shape of the distribution. Further, at least eight of nine values, or 88.9%, will lie between plus three standard deviations and minus three standard deviations of the mean. At least 24 of the 25 values, or 96 percent, will lie between plus and minus five standard deviations of the mean.

Chebyshev's theorem states:

CHEBYSHEV'S THEOREM For any set of observations (sample or population), the proportion of the values that lie within k standard deviations of the mean is at least $1 - 1/k^2$, where k is any constant greater than 1.

EXAMPLE

The average biweekly amount contributed by the Dupree Paint employees to the company's profit-sharing plan is $51.54, and the standard deviation is $7.51. At least what percent of the contributions lie within plus and minus 3.5 standard deviations of the mean?

Solution

About 92 percent, found by $1 - 1/k^2 = 1 - 1/(3.5)^2 = 1 - 1/12.25 = 0.92$

The Empirical Rule

Chebyshev's theorem is concerned with any set of values; that is, the distribution of values can have any shape. However, for a symmetrical, bell-shaped distribution such as the one in Chart 3–7, we can be more precise in describing the dispersion about the mean. These relationships involving the standard deviation and the mean are called the **Empirical Rule,** or the **Normal Rule.**

Empirical Rule applies only to symmetrical, bell-shaped distributions.

EMPIRICAL RULE For a symmetrical, bell-shaped frequency distribution, approximately 68 percent of the observations will lie within plus and minus one standard deviation of the mean; about 95 percent of the observations will lie within plus and minus two standard deviations of the mean; and practically all (99.7 percent) will lie within plus and minus three standard deviations of the mean.

These relationships are portrayed graphically in Chart 3–7 for a bell-shaped distribution with a mean of 100 and a standard deviation of 10.

It has been noted that if a distribution is symmetrical and bell-shaped, practically all of the observations lie between the mean plus and minus three standard deviations. Thus, if

Chapter 3

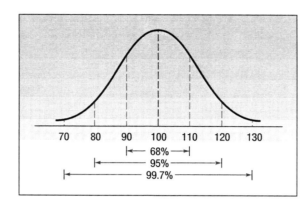

CHART 3–7 A Symmetrical, Bell-Shaped Curve Showing the Relationships between the Standard Deviation and the Observations

$\bar{X} = 100$ and $s = 10$, practically all the observations lie between $100 + 3(10)$ and $100 - 3(10)$, or 70 and 130. The range is therefore 60, found by $130 - 70$.

Conversely, if we know that the range is 60, we can approximate the standard deviation by dividing the range by 6. For this illustration: range $\div 6 = 60 \div 6 = 10$, the standard deviation.

It should be noted that the Empirical Rule does not violate Chebyshev's theorem by increasing the percentages.

EXAMPLE

A sample of the monthly amounts spent for food by a senior citizen living alone approximates a symmetrical, bell-shaped distribution. The sample mean is $225; the standard deviation is $20. Using the Empirical Rule:

1. About 68 percent of the monthly food expenditures are between what two amounts?
2. About 95 percent of the monthly food expenditures are between what two amounts?
3. Almost all of the monthly expenditures are between what two amounts?

Solution

1. About 68 percent are between $205 and $245, found by $\bar{X} \pm 1s = \$225 \pm 1(\$20)$.
2. About 95 percent are between $185 and $265, found by $\bar{X} \pm 2s = \$225 \pm 2(\$20)$.
3. Almost all (99.7 percent) are between $165 and $285, found by $\bar{X} \pm 3s = \$225 \pm 3(\$20)$.

Self-Review 3–9

The Superior Metal Company is one of several domestic manufacturers of steel pipe. The quality control department sampled 600 10 m lengths. At a point 1 m from the end of the pipe they measured the outside diameter. The mean was 1.2 m and the standard deviation 0.1 m.

(a) If the shape of the distribution is not known, at least what percent of the observations will lie between 1.05 m and 1.35 m?
(b) If we assume that the distribution of diameters is symmetrical and bell-shaped, about 68 percent of the observations will be between what two values?
(c) If we assume that the distribution of diameters is symmetrical and bell-shaped, about 95 percent of the observations will be between what two values?

Exercises

47. According to Chebyshev's theorem, at least what percent of any set of observations will be within 1.8 standard deviations of the mean?

Describing Data: Numerical Measures

48. The mean income of a group of sample observations is $500; the standard deviation is $40. According to Chebyshev's theorem at least what percent of the incomes will lie between $400 and $600?

49. The distribution of the weights of a sample of 1400 cargo containers is somewhat normally distributed. Based on the Empirical Rule, what percent of the weights will lie
  a. Between $\bar{X} - 2s$ and $\bar{X} + 2s$?
  b. Between $\bar{X}$ and $\bar{X} + 2s$? Below $\bar{X} - 2s$?

50. The following figure portrays the appearance of a distribution of efficiency ratings for employees of Nale Nail Works, Inc.

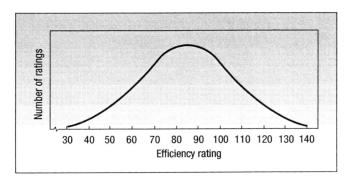

  a. Estimate the mean efficiency rating.
  b. Estimate the standard deviation to the nearest whole number.
  c. About 68 percent of the efficiency ratings are between what two values?
  d. About 95 percent of the efficiency ratings are between what two values?

Relative Dispersion

A direct comparison of two or more measures of dispersion—say, the standard deviation for a distribution of annual incomes and the standard deviation of a distribution of absenteeism for this same group of employees—is impossible. Can we say that the standard deviation of $1200 for the income distribution is greater than the standard deviation of 4.5 days for the distribution of absenteeism? Obviously not, because we cannot directly compare dollars and days absent from work. In order to make a meaningful comparison of the dispersion in incomes and absenteeism, we need to convert each of these measures to a *relative value*—that is, a percent. Karl Pearson (1857–1936), who contributed significantly to the science of statistics, developed a relative measure called the **coefficient of variation** (CV). It is a very useful measure when:

When to use CV

1. The data are in different units (such as dollars and days absent).
2. The data are in the same units, but the means are far apart (such as the incomes of the top executives and the incomes of the unskilled employees).

COEFFICIENT OF VARIATION The ratio of the standard deviation to the arithmetic mean, expressed as a percent.

In terms of a formula for a sample:

COEFFICIENT OF VARIATION $\quad CV = \dfrac{s}{\bar{X}}(100) \leftarrow$ Multiplying by 100 converts the decimal to a percent $\quad$ [3–13]

EXAMPLE

A study of the amount of bonus paid and the years of service of employees at Sea Pro Marine, Inc., resulted in these statistics: The mean bonus paid was $200; the standard deviation was $40. The mean number of years of service was 20 years; the standard deviation was 2 years. Compare the relative dispersion in the two distributions using the coefficient of variation.

Chapter 3

Solution

The distributions are in different units (dollars and years of service). Therefore, they are converted to coefficients of variation.

For the bonus paid:

$$CV = \frac{s}{\bar{X}}(100)$$

$$= \frac{\$40}{\$200}(100)$$

$$= 20 \text{ percent}$$

For years of service:

$$CV = \frac{s}{\bar{X}}(100)$$

$$= \frac{2}{20}(100)$$

$$= 10 \text{ percent}$$

Interpreting, there is more dispersion relative to the mean in the distribution of bonus paid compared with the distribution of years of service (because 20 percent > 10 percent).

The same procedure is used when the data are in the same units but the means are far apart. (See the following example.)

EXAMPLE

The variation in the annual incomes of executives at Nash-Rambler Products, Inc. is to be compared with the variation in incomes of unskilled employees. For a sample of executives, $\bar{X} = \$500\,000$ and $s = \$50\,000$. For a sample of unskilled employees, $\bar{X} = \$32\,000$ and $s = \$3200$. We are tempted to say that there is more dispersion in the annual incomes of the executives because $\$50\,000 > \$3200$. The means are so far apart, however, that we need to convert the statistics to coefficients of variation to make a meaningful comparison of the variations in annual incomes.

Solution

For the executives:

$$CV = \frac{s}{\bar{X}}(100)$$

$$= \frac{\$50\,000}{\$500\,000}(100)$$

$$= 10 \text{ percent}$$

For the unskilled employees:

$$CV = \frac{s}{\bar{X}}(100)$$

$$= \frac{\$3200}{\$32\,000}(100)$$

$$= 10 \text{ percent}$$

There is no difference in the relative dispersion of the two groups.

Self-Review 3–10

A large group of Air Force inductees was given two experimental tests—a mechanical aptitude test and a finger dexterity test. The arithmetic mean score on the mechanical aptitude test was 200, with a standard deviation of 10. The mean and standard deviation for the finger dexterity test were: $\bar{X} = 30$, $s = 6$. Compare the relative dispersion in the two groups.

Exercises

51. For a sample of students in a college studying Business Administration, the mean grade point average is 3.10 with a standard deviation of 0.25. Compute the coefficient of variation.
52. Skipjack Airlines is studying the mass of luggage for each passenger. For a large group of domestic passengers, the mean is 21 kg with a standard deviation of 5 kg. For a large group of overseas passengers, the mean is 35 kg and the standard deviation is 7 kg. Compute the relative dispersion of each group. Comment on the difference in relative dispersion.
53. The research analyst for the Sidde Financial stock brokerage firm wants to compare the dispersion in the price-earnings ratios for a group of common stocks with the dispersion of their return on investment. For the price-earnings ratios, the mean is 4.9 and the standard deviation 1.8. The mean return on investment is 10 percent and the standard deviation 5.2 percent.
    a. Why should the coefficient of variation be used to compare the dispersion?
    b. Compare the relative dispersion for the price-earnings ratios and return on investment.

54. The spread in the annual prices of stocks selling for under $10 and the spread in prices of those selling for over $60 are to be compared. The mean price of the stocks selling for under $10 is $5.25 and the standard deviation $1.52. The mean price of those stocks selling for over $60 is $92.50 and the standard deviation $5.28.
  a. Why should the coefficient of variation be used to compare the dispersion in the prices?
  b. Compute the coefficients of variation. What is your conclusion?

Skewness

In this chapter we have described measures of location of a set of observations by reporting the mean, median, and mode. We have also described measures that show the amount of spread or variation in a set of data, such as the range and the standard deviation.

Another characteristic of a set of data is the shape. There are four shapes commonly observed: symmetric, positively skewed, negatively skewed, and bimodal. In a **symmetric** set of observations the mean and median are equal and the data values are evenly spread around these values. The data values below the mean and median are a mirror image of those above. A set of values is **skewed to the right** or **positively skewed** if there is a single peak and the values extend much further to the right of the peak than to the left of the peak. In this case the mean is larger than the median. In a **negatively skewed** distribution there is a single peak but the observations extend further to the left, in the negative direction, than to the right. In a negatively skewed distribution the mean is smaller than the median. Positively skewed distributions are more common. Salaries often follow this pattern. Think of the salaries of those employed in a small company of about 100 people. The president and a few top executives would have very large salaries relative to the other workers and hence the distribution of salaries would exhibit positive skewness. A **bimodal distribution** will have two or more peaks. This is often the case when the values are from two or more populations. This information is summarized in Chart 3–8.

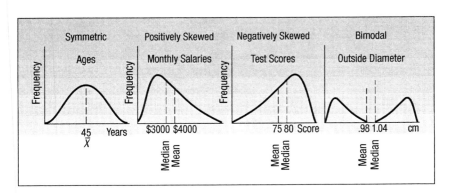

CHART 3–8 Shapes of Frequency Distributions

There are several formulas in the statistical literature used to calculate skewness. The simplest, developed by Professor Karl Pearson, is based on the difference between the mean and the median.

PEARSON'S COEFFICIENT OF SKEWNESS $$sk = \frac{3(\bar{X} - \text{Median})}{s}$$ [3–14]

Using this relationship the coefficient of skewness can range from −3 up to 3. A value near −3, such as −2.57, indicates considerable negative skewness. A value such as 1.63 indicates moderate positive skewness. A value of 0, which will occur when the mean and median are equal, indicates the distribution is symmetrical and that there is no skewness present.

An example will illustrate the idea of skewness.

EXAMPLE

Following are the earnings per share, in dollars, for a sample of 15 software companies for the year 2002. The earnings per share are arranged from smallest to largest.

0.09	0.13	0.41	0.51	1.12	1.20	1.49	3.18
3.50	6.36	7.83	8.92	10.13	12.99	16.40	

Compute the mean, median, and standard deviation. Find the coefficient of skewness using Pearson's estimate. What is your conclusion regarding the shape of the distribution?

Solution

These are sample data, so we use formula (3–2) to determine the mean

$$\bar{X} = \frac{\Sigma X}{n} = \frac{\$74.26}{15} = \$4.95$$

The median is the middle value in a set of data, arranged from smallest to largest. In this case the middle value is $3.18, so the median earnings per share is $3.18.

We use formula (3–12) on page 86 to determine the sample standard deviation.

$$s = \sqrt{\frac{\Sigma X^2 - \frac{(\Sigma X)^2}{n}}{n-1}} = \sqrt{\frac{749.372 - \frac{(74.26)^2}{15}}{15-1}} = \$5.22$$

Pearson's coefficient of skewness is 1.017, found by

$$sk = \frac{3(\bar{X} - \text{Median})}{n-1} = \frac{3(\$4.95 - \$3.18)}{\$5.22} = 1.017$$

This indicates there is moderate positive skewness in the earnings per share data.

The MegaStat commands to answer this example follow. Note that computer calculations may be slightly different due to rounding.

Self-Review 3–11

A sample of five data entry clerks employed in the customer service department of a large pharmaceutical distribution company revised the following number of records last hour: 73, 98, 60, 92, and 84.

(a) Find the mean, median, and the standard deviation.
(b) Compute the coefficient of skewness using Pearson's method.
(c) What is your conclusion regarding the skewness of the data?

Describing Data: Numerical Measures

Exercises

For Exercises 55–58, do the following:

a. Determine the mean, median, and the standard deviation.
b. Determine the coefficient of skewness using Pearson's method.

55. The following values are the starting salaries, in thousands of dollars, for a sample of five accounting graduates who accepted positions in public accounting last year.

36.0	26.0	33.0	28.0	31.0

56. Listed below are the salaries, in thousands of dollars, for a sample of 15 chief financial officers in the electronics industry.

516.0	548.0	566.0	534.0	586.0	529.0
546.0	523.0	538.0	523.0	551.0	552.0
486.0	558.0	574.0			

57. Listed below are the commissions earned, in thousands of dollars, last year by the sales representatives at the Furniture Patch, Inc.

3.9	5.7	7.3	10.6	13.0	13.6	15.1	15.8	17.1
17.4	17.6	22.3	38.6	43.2	87.7			

58. Listed below are the salaries for the New York Yankees for the year 2000. The salary information is reported in millions of dollars.

9.86	9.50	8.25	6.25	6.00	5.95
5.25	5.00	4.33	4.30	4.25	3.40
3.13	2.02	2.00	1.90	1.85	1.82
0.80	0.38	0.35	0.35	0.20	0.20
0.20	0.20	0.20	0.20	0.20	

Other Measures of Dispersion

The standard deviation is the most widely used measure of dispersion. However, there are other ways of describing the variation or spread in a set of data. One method is to determine the *location* of values that divide a set of observations into equal parts. These measures include *quartiles, deciles,* and *percentiles.*

Quartiles divide a set of observations into four equal parts. To explain further, think of any set of values arranged from smallest to largest. Earlier in this chapter we called the middle value of a set of data arranged from smallest to largest the median. That is, 50 percent of the observations are larger than the median and 50 percent are smaller. The median is a measure of location because it pinpoints the centre of the data. In a similar fashion quartiles divide a set of observations into four equal parts. The first quartile, usually labeled Q_1, is the value below which 25 percent of the observations occur, and the third quartile, usually labeled Q_3, is the value below which 75 percent of the observations occur. Logically, Q_2 is the median. The values corresponding to Q_1, Q_2, and Q_3 divide a set of data into four equal parts. Q_1 can be thought of as the "median" of the lower half of the data and Q_3 the "median" of the upper half of the data.

In a similar fashion deciles divide a set of observations into 10 equal parts and percentiles into 100 equal parts. So if you found that your GPA was in the 8th decile at your school, you could conclude that 80 percent of the students had a GPA lower than yours and 20 percent had a higher GPA. A GPA in the 33rd percentile means that 33 percent of the students have a lower GPA and 67 percent have a higher GPA. Percentile scores are frequently used to report results on such national standardized tests as the SAT, ACT, GMAT (used to judge entry into many Master of Business Administration programs), and LSAT (used to judge entry into law school).

Quartiles, Deciles, and Percentiles

To formalize the computational procedure, let L_p refer to the location of a desired percentile. So if we wanted to find the 33rd percentile we would use L_{33} and if we wanted the median, the 50th percentile, then L_{50}. The number of observations is n, so if we want to locate the middle observation, its position is at $(n + 1)/2$, or we could write this as $(n + 1)(P/100)$, where P is the desired percentile.

LOCATION OF A PERCENTILE $\qquad L_p = (n + 1)\dfrac{P}{100} \qquad$ [3–15]

An example will help to explain further.

EXAMPLE

Listed below are the commissions earned, in dollars, last month by a sample of 15 brokers at Salomon Smith Barney's office.

2038	1758	1721	1637	2097	2047	2205	1787	2287
1940	2311	2054	2406	1471	1460			

Locate the median, the first quartile, and the third quartile for the commissions earned.

Solution

The first step is to organize the data from the smallest commission to the largest.

1460	1471	1637	1721	1758	1787	1940	2038
2047	2054	2097	2205	2287	2311	2406	

The median value is the observation in the centre. The centre value or L_{50} is located at $(n + 1)(50/100)$, where n is the number of observations. In this case that is position number 8, found by $(15 + 1)(50/100)$. The eighth largest commission is $2038. So we conclude this is the median and that half the brokers earned commissions more than $2038 and half earned less than $2038.

Recall the definition of a quartile. Quartiles divide a set of observations into four equal parts. Hence 25 percent of the observations will be less than the first quartile. Seventy-five percent of the observations will be less than the third quartile. To locate the first quartile, we use formula (3–16), where $n = 15$ and $P = 25$:

$$L_{25} = (n + 1)\dfrac{P}{100} = (15 + 1)\dfrac{25}{100} = 4$$

and to locate the third quartile, $n = 15$ and $P = 75$:

$$L_{75} = (n + 1)\dfrac{P}{100} = (15 + 1)\dfrac{75}{100} = 12$$

Therefore, the first and third quartile values are located at positions 4 and 12. The fourth value in the ordered array is $1721 and the twelfth is $2205. These are the first and third quartiles, respectively.

In the above example the location formula yielded a whole number result. That is, we wanted to find the first quartile and there were 15 observations, so the location formula indicated we should find the fourth ordered value. What if there were 20 observations in the sample, that is $n = 20$, and we wanted to locate the first quartile? From the location formula (3–16):

$$L_{25} = (n + 1)\dfrac{P}{100} = (20 + 1)\dfrac{25}{100} = 5.25$$

We would locate the fifth value in the ordered array and then move .25 of the distance between the fifth and sixth values and report that as the first quartile. Like the median, the quartile does not need to be one of the actual values in the data set.

Describing Data: Numerical Measures

To explain further, suppose a data set contained the six values: 91, 75, 61, 101, 43, and 104. We want to locate the first quartile. We order the values from smallest to largest: 43, 61, 75, 91, 101, and 104. The first quartile is located at

$$L_{25} = (n+1)\frac{P}{100} = (6+1)\frac{25}{100} = 1.75$$

The position formula tells us that the first quartile is located between the first and the second value and that it is .75 of the distance between the first and the second values. The first value is 43 and the second is 61. So the distance between these two values is 18. To locate the first quartile, we need to move .75 of the distance between the first and second values, so .75(18) = 13.5. To complete the procedure, we add 13.5 to the first value and report that the first quartile is 56.5.

We can extend the idea to include both deciles and percentiles. If we wanted to locate the 23rd percentile in a sample of 80 observations, we would look for the 18.63 position.

$$L_{25} = (n+1)\frac{P}{100} = (80+1)\frac{23}{100} = 18.63$$

To find the value corresponding to the 23rd percentile, we would locate the 18th value and the 19th value and determine the distance between the two values. Next, we would multiply this difference by 0.63 and add the result to the smaller value. The result would be the 23rd percentile.

The Excel commands to calculate quartiles and percentiles follow. The method used is not as precise as that described above, but the result is a good approximation. Using the real estate example data (Table 02-1), we will calculate the 1st quartile (25th percentile) and the 56th percentile. Use the Paste Function, Statistical category, Quartile and then Percentile. The result appears in the dialogue boxes, and by clicking OK, the result will appear in the worksheet.

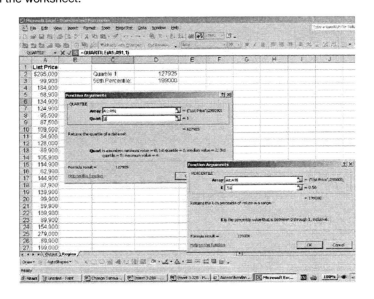

Self-Review 3-12

The quality control department of the Plainsville Peanut Company is responsible for checking the mass of the 500 g jar of peanut butter. The masses of a sample of nine jars produced last hour are:

| 490 | 495 | 496 | 498 | 500 | 500 | 501 | 504 | 505 |

(a) What is the median mass?
(b) Determine the masses corresponding to the first and third quartiles.

Exercises

59. Determine the median and the values corresponding to the first and third quartiles in the following data.

| 46 | 47 | 49 | 49 | 51 | 53 | 54 | 54 | 55 | 55 | 59 |

60. Determine the median and the values corresponding to the first and third quartiles in the following data.

| 5.24 | 6.02 | 6.67 | 7.30 | 7.59 | 7.99 | 8.03 | 8.35 | 8.81 | 9.45 |
| 9.61 | 10.37 | 10.39 | 11.86 | 12.22 | 12.71 | 13.07 | 13.59 | 13.89 | 15.42 |

61. The Thomas Supply Company, Inc. is a distributor of small electrical motors. As with any business, the length of time customers take to pay their invoices is important. Listed below, arranged from smallest to largest, is the time, in days, for a sample of The Thomas Supply Company, Inc. invoices.

| 13 | 13 | 13 | 20 | 26 | 27 | 31 | 34 | 34 | 34 | 35 | 35 | 36 | 37 | 38 |
| 41 | 41 | 41 | 45 | 47 | 47 | 47 | 50 | 51 | 53 | 54 | 56 | 62 | 67 | 82 |

   a. Determine the first and third quartiles.
   b. Determine the second decile and the eighth decile.
   c. Determine the 67th percentile.

62. Kevin Horn is the national sales manager for National Textbooks, Inc. He has a sales staff of 40 who visit college and university professors. Each Saturday morning he requires his sales staff to send him a report. This report includes, among other things, the number of professors visited during the previous week. Listed below, ordered from smallest to largest, are the number of visits last week.

| 38 | 40 | 41 | 45 | 48 | 48 | 50 | 50 | 51 | 51 | 52 | 53 | 54 | 55 | 55 | 55 | 56 | 56 | 57 |
| 59 | 59 | 59 | 62 | 62 | 62 | 63 | 64 | 65 | 66 | 66 | 67 | 67 | 69 | 69 | 71 | 77 | 78 | 79 | 79 |

   a. Determine the median number of calls.
   b. Determine the first and third quartiles.
   c. Determine the first decile and the ninth decile.
   d. Determine the 33rd percentile.

Box Plots

A **box plot** is a graphical display, based on quartiles, that helps us picture a set of data. To construct a box plot, we need only five statistics: the minimum value, Q_1 (the first quartile), the median, Q_3 (the third quartile), and the maximum value. An example will help to explain.

EXAMPLE

Alexander's Pizza offers free delivery of its pizza within 15 km. Alex, the owner, wants some information on the time it takes for delivery. How long does a typical delivery take? Within what range of times will most deliveries be completed? For a sample of 20 deliveries, he determined the following information:

$$\text{Minimum value} = 13 \text{ minutes}$$
$$Q_1 = 15 \text{ minutes}$$
$$\text{Median} = 18 \text{ minutes}$$
$$Q_3 = 22 \text{ minutes}$$
$$\text{Maximum value} = 30 \text{ minutes}$$

Develop a box plot for the delivery times. What conclusions can you make about the delivery times?

Solution

The first step in drawing a box plot is to create an appropriate scale along the horizontal axis. Next, we draw a box that starts at Q_1 (15 minutes) and ends at Q_3 (22 minutes). Inside

Describing Data: Numerical Measures

the box we place a vertical line to represent the median (18 minutes). Finally, we extend horizontal lines from the box out to the minimum value (13 minutes) and the maximum value (30 minutes). These horizontal lines outside of the box are sometimes called "whiskers" because they look a bit like a cat's whiskers.

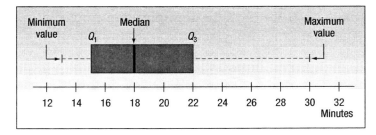

The box plot shows that the middle 50 percent of the deliveries take between 15 minutes and 22 minutes. The distance between the ends of the box, 7 minutes, is the **interquartile range.** The interquartile range is the distance between the first and the third quartile. It shows the spread or dispersion of the majority of deliveries.

The box plot also reveals that the distribution of delivery times is positively skewed. How do we know this? In this case there are actually two pieces of information that suggest that the distribution is positively skewed. First, the dashed line to the right of the box from 22 minutes (Q_3) to the maximum time of 30 minutes is longer than the dashed line from the left of 15 minutes (Q_1) to the minimum value of 13 minutes. To put it another way, the 25 percent of the data larger than the third quartile is more spread out than the 25 percent less than the first quartile. A second indication of positive skewness is that the median is not in the centre of the box. The distance from the first quartile to the median is smaller than the distance from the median to the third quartile. We know that the number of delivery times between 15 minutes and 18 minutes is the same as the number of delivery times between 18 minutes and 22 minutes.

EXAMPLE

Refer to the Real Estate data in Table 2–1. Develop a box plot of the data. What can we conclude about the distribution of the list prices?

Solution

MegaStat was used to develop the following chart. The commands follow.

We can conclude that the median list price is about $190 000, that about 25 percent of the list prices are below $125 000 and that about 25 percent are above $240 000. About 50 percent of the list prices are between $125 000 and $240 000. The distribution is positively skewed because the solid line above $240 000 is somewhat longer than the line below $190 000.

There are 3 small circles above the $400 000 list price. The circles represent outliers. An **outlier** is a value that is more than 1.5 times the interquartile range larger than Q_3, or smaller than Q_1. In this example, $Q_1 = \$127\,925$, $Q_3 = 232\,225$, and the interquartile range $= 104\,300\ (232\,225 - 127\,925)$. So a high outlier would be a value larger than $388\,675$, found by

$$\text{Outlier} > Q_3 + 1.5(Q_3 - Q_1) = \$232\,225 + 1.5(\$104\,300) = \$388\,675$$

A value less than $\$ - 28\,525$ is also an outlier, but would be described as a low outlier. Since there are no list prices less than $50 000, there are no low outliers for this data.

$$\text{Outlier} < Q_1 - 1.5(Q_3 - Q_1) = \$127\,925 - 1.5(\$104\,300) = \$ - 28\,525$$

Self-Review 3–13

The following box plot is given.

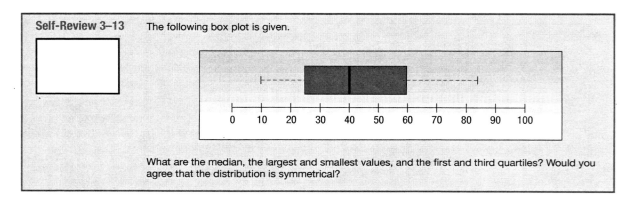

What are the median, the largest and smallest values, and the first and third quartiles? Would you agree that the distribution is symmetrical?

Exercises

63. Refer to the box plot below.

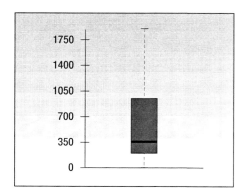

a. Estimate the median.
b. Estimate the first and third quartiles.
c. Determine the interquartile range.
d. Beyond what point is a value considered an outlier?
e. Identify any outliers and estimate their value.
f. Is the distribution symmetrical or positively or negatively skewed?

Describing Data: Numerical Measures

64. Refer to the following box plot.

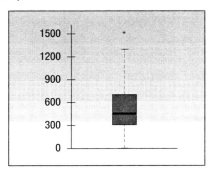

   a. Estimate the median.
   b. Estimate the first and third quartiles.
   c. Determine the interquartile range.
   d. Beyond what point is a value considered an outlier?
   e. Identify any outliers and estimate their value.
   f. Is the distribution symmetrical or positively or negatively skewed?

65. In a study of the fuel efficiency of model year 2002 automobiles, the mean efficiency was 12 km/L and the median was 11.7 km/L. The smallest value in the study was 5.5 km/L, and the largest was 22 km/L. The first and third quartiles were 7.8 km/L and 15.5 km/L, respectively. Develop a box plot and comment on the distribution. Is it a symmetric distribution?

66. A sample of 28 hospitals revealed the following daily charges, in dollars, for a semi-private room. For convenience the data are ordered from smallest to largest. Construct a box plot to represent the data. Comment on the distribution. Be sure to identify the first and third quartiles and the median.

116	121	157	192	207	209	209
229	232	236	236	239	243	246
260	264	276	281	283	289	296
307	309	312	317	324	341	353

The Mean and Standard Deviation of Grouped Data

In most instances measures of location, such as the mean, and measures of variability, such as the standard deviation, are determined by using the individual values. Excel makes it easy to calculate these values, even for large data sets. However, sometimes we are only given the frequency distribution and wish to estimate the mean or standard deviation. In the following discussion we show how we can estimate the mean and standard deviation from data organized into a frequency distribution. We should stress that a mean or a standard deviation from grouped data is an *estimate* of the corresponding actual values.

The Arithmetic Mean

To approximate the arithmetic mean of data organized into a frequency distribution, we begin by assuming the observations in each class are represented by the *midpoint* of the class. The mean of a sample of data organized in a frequency distribution is computed by:

ARITHMETIC MEAN OF GROUPED DATA $\bar{X} = \dfrac{\Sigma fX}{n}$ [3–16]

where:
  $\bar{X}$ is the designation for the sample mean.
  X is the midpoint of each class.
  f is the frequency in each class.
  fX is the frequency in each class times the midpoint of the class.
  ΣfX is the sum of these products.
  n is the total number of frequencies.

EXAMPLE

The computations for the arithmetic mean of data grouped into a frequency distribution will be shown based on the real estate, Regina and surrounding area data. Recall in Chapter 2, in Table 2–4 we constructed a frequency distribution for the list prices. The information is repeated below. Determine the arithmetic mean list price.

List Price ($ thousands)	Frequency
50 to under 100	14
100 to under 150	23
150 to under 200	16
200 to under 250	18
250 to under 300	8
300 to under 350	5
350 to under 400	4
400 to under 450	2
Total	90

Solution

The mean vehicle selling price can be estimated from data grouped into a frequency distribution. To find the estimated mean, assume the midpoint of each class is representative of the data values in that class. Recall that the midpoint of a class is halfway between the upper and the lower class limits. To find the midpoint of a particular class, we add the upper and the lower class limits and divide by 2. Hence, the midpoint of the first class is $75, found by ($50 + $100)/2. We assume that the value of $75 is representative of the 14 values in that class. To put it another way, we assume the sum of the 14 values is $1050, found by 14($75). We continue the process of multiplying the class midpoint by the class frequency for each class and then sum these products. The results are summarized in Table 3–1.

TABLE 3–1 List Prices, Regina & Surrounding Area

List Price ($ thousands)	Frequency f	Midpoint ($) X	fX ($)
50 to under 100	14	75	1050
100 to under 150	23	125	2875
150 to under 200	16	175	2800
200 to under 250	18	225	4050
250 to under 300	8	275	2200
300 to under 350	5	325	1625
350 to under 400	4	375	1500
400 to under 450	2	425	850
Total	90		16 950

Solving for the arithmetic mean using formula (3–17), we get:

$$\bar{X} = \frac{\Sigma fX}{n} = \frac{\$16\,950}{90} = \$188.33 \text{ (thousands)}$$

So we conclude that the mean list price is about $188 333.33

Standard Deviation

Recall that for *ungrouped* data, one formula for the sample standard deviation is:

$$s = \sqrt{\frac{\Sigma X^2 - \frac{(\Sigma X)^2}{n}}{n-1}}$$

Describing Data: Numerical Measures

If the data of interest are in *grouped* form (in a frequency distribution), the sample standard deviation can be approximated by substituting ΣfX^2 for ΣX^2 and ΣfX for ΣX. The formula for the *sample standard deviation* then converts to:

STANDARD DEVIATION; GROUPED DATA
$$s = \sqrt{\frac{\Sigma fX^2 - \frac{(\Sigma fX)^2}{n}}{n-1}}$$ [3–17]

where:
- s is the symbol for the sample standard deviation.
- X is the midpoint of a class.
- f is the class frequency.
- n is the total number of sample observations.

EXAMPLE

Refer to the frequency distribution for the real estate data reported in Table 3–1. Compute the standard deviation of the list prices.

Solution

Following the same practice used earlier for computing the mean of data grouped into a frequency distribution, X represents the midpoint of each class.

TABLE 3–1 List Prices, Regina & Surrounding Area

List Price ($ thousands)	Frequency f	Midpoint ($) X	fX ($)	fX²
50 to under 100	14	75	1050	78 750
100 to under 150	23	125	2875	359 375
150 to under 200	16	175	2800	490 000
200 to under 250	18	225	4050	911 250
250 to under 300	8	275	2200	605 000
300 to under 350	5	325	1625	528 125
350 to under 400	4	375	1500	562 500
400 to under 450	2	425	850	361 250
Total	90		16 950	3 896 250

To find the standard deviation:

Step 1: Each class frequency is multiplied by its class midpoint. That is, multiply f times X. Thus, for the first class $14 \times \$75 = \$1050$, for the second class $fX = 23 \times \$125 = \$2875$, and so on.

Step 2: Calculate fX^2. This could be written $fX \times X$. For the first class it would be $1050 \times 14 = 78\,750$, for the second class it would be $2875 \times 23 = 359\,375$, and so on.

Step 3: Sum the fX and fX^2 columns. The totals are $\$16\,950$ and $\$3\,896\,250$, respectively. We have omitted the units involved with the fX^2 column, but it is "dollars squared."

To find the standard deviation we insert these values in formula (3–18).

$$s = \sqrt{\frac{\Sigma fX^2 - \frac{(fX)^2}{n}}{n-1}} = \sqrt{\frac{3\,896\,250 - \frac{(16\,950)^2}{90}}{90-1}} = 88.939$$

The mean and standard deviation calculated from data grouped into a frequency distribution are usually close to the values calculated from raw data. The grouping results in some loss of information. For the list problem the mean price reported in the Excel output is $192\,712$ and the standard deviation is $86\,881$. The respective values estimated from data grouped into a frequency distribution are $188\,333$ and $88\,939$. The difference in the means is 4379 or about 2.8 percent. The standard deviations differ by 2058 or 2.4 percent. Based on the percentage difference, the estimates are very close to the actual values.

Self-Review 3–14

The net incomes of a sample of large importers of antiques were organized into the following table:

Net Income ($ millions)	Number of Importers
2 to under 6	1
6 to under 10	4
10 to under 14	10
14 to under 18	3
18 to under 22	2

(a) What is the table called?
(b) Based on the distribution, what is the estimate of the arithmetic mean net income?
(c) Based on the distribution, what is the estimate of the standard deviation?

Exercises

67. When we compute the mean of a frequency distribution, why do we refer to this as an *estimated* mean?

68. Determine the mean and the standard deviation of the following frequency distribution.

Class	Frequency
0 to under 5	2
5 to under 10	7
10 to under 15	12
15 to under 20	6
20 to under 25	3

69. Determine the mean and the standard deviation of the following frequency distribution.

Class	Frequency
20 to under 30	7
30 to under 40	12
40 to under 50	21
50 to under 60	18
60 to under 70	12

70. SCCoast, an Internet provider, developed the following frequency distribution on the age of Internet users. Find the mean and the standard deviation.

Age (years)	Frequency
10 to under 20	3
20 to under 30	7
30 to under 40	18
40 to under 50	20
50 to under 60	12

71. The following frequency distribution reports the amount, in thousands of dollars, owed by a sample of 50 public accounting firms. Find the mean and the standard deviation.

Amount ($ thousands)	Frequency
20 to under 30	1
30 to under 40	15
40 to under 50	22
50 to under 60	8
60 to under 70	4

Describing Data: Numerical Measures

72. Advertising expenses are a significant component of the cost of goods sold. Listed below is a frequency distribution showing the advertising expenditures for 60 manufacturing companies. Estimate the mean and the standard deviation of advertising expense.

Advertising Expenditure ($ millions)	Number of Companies
25 to under 35	5
35 to under 45	10
45 to under 55	21
55 to under 65	16
65 to under 75	8
Total	60

Chapter Outline

I. A measure of location is a value used to describe the centre of a set of data.

   A. The arithmetic mean is the most widely reported measure of location.
   1. It is calculated by adding the values of the observations and dividing by the total number of observations.
      a. The formula for a population mean of ungrouped or raw data is:
      $$\mu = \frac{\Sigma X}{N} \qquad [3\text{--}1]$$
      b. The formula for the mean of a sample is
      $$\bar{X} = \frac{\Sigma X}{n} \qquad [3\text{--}2]$$
      c. The formula for the sample mean of data in a frequency distribution is
      $$\bar{X} = \frac{\Sigma fX}{n} \qquad [3\text{--}16]$$
   2. The major characteristics of the arithmetic mean are:
      a. At least the interval scale of measurement is required.
      b. All the data values are used in the calculation.
      c. A set of data has only one mean. That is, it is unique.
      d. The sum of the deviations from the mean equals 0.

   B. The weighted mean is found by multiplying each observation by its corresponding weight.
   1. The formula for determining the weighted mean is:
   $$\bar{X}_W = \frac{w_1 X_1 + w_2 X_2 + w_3 X_3 + \cdots + w_n X_n}{w_1 + w_2 + w_3 + \cdots + w_n} \qquad [3\text{--}3]$$
   2. It is a special case of the arithmetic mean.

   C. The median is the value in the middle of a set of ordered data.
   1. To find the median, sort the observations from smallest to largest and identify the middle value.
   2. The major characteristics of the median are:
      a. At least the ordinal scale of measurement is required.
      b. It is not influenced by extreme values.
      c. Fifty percent of the observations are larger than the median.
      d. It is unique to a set of data.

   D. The mode is the value that occurs most often in a set of data.
   1. The mode can be found for nominal-level data.
   2. A set of data can have more than one mode.

Chapter 3

E. The geometric mean is the nth root of the product of n values.
   1. The formula for the geometric mean is:

   $$GM = \sqrt[n]{(X_1)(X_2)(X_3)\cdots(X_n)} \qquad [3\text{–}4]$$

   2. The geometric mean is also used to find the rate of change from one period to another.

   $$GM = \sqrt[n]{\frac{\text{Value at end of period}}{\text{Value at beginning of period}}} - 1 \qquad [3\text{–}5]$$

   3. The geometric mean is always equal to or less than the arithmetic mean.

II. The dispersion is the variation or spread in a set of data.

   A. The range is the difference between the largest and the smallest value in a set of data.
      1. The formula for the range is:

      $$\text{Range} = \text{Highest value} - \text{Lowest value} \qquad [3\text{–}6]$$

      2. The major characteristics of the range are:
         a. Only two values are used in its calculation.
         b. It is influenced by extreme values.
         c. It is easy to compute and to understand.

   B. The mean absolute deviation is the sum of the absolute deviations from the mean divided by the number of observations.
      1. The formula for computing the mean absolute deviation is

      $$MD = \frac{\Sigma |X - \bar{X}|}{n} \qquad [3\text{–}7]$$

      2. The major characteristics of the mean absolute deviation are:
         a. It is not unduly influenced by large or small values.
         b. All observations are used in the calculation.
         c. The absolute values are somewhat difficult to work with.

   C. The variance is the mean of the squared deviations from the arithmetic mean.
      1. The formula for the population variance is:

      $$\sigma^2 = \frac{\Sigma(X - \mu)^2}{N} \qquad [3\text{–}8]$$

      2. The formula for the sample variance is:

      $$s^2 = \frac{\Sigma(X - \bar{X})^2}{n - 1} \qquad [3\text{–}10]$$

      3. The major characteristics of the variance are:
         a. All observations are used in the calculation.
         b. It is not unduly influenced by extreme observations.
         c. The units are somewhat difficult to work with; they are the original units squared.

   D. The standard deviation is the square root of the variance.
      1. The major characteristics of the standard deviation are:
         a. It is in the same units as the original data.
         b. It is the square root of the average squared deviation from the mean.
         c. It cannot be negative.
         d. It is the most widely reported measure of dispersion.
      2. The formula for the sample standard deviation is:

      $$s = \sqrt{\frac{\Sigma X^2 - \frac{(\Sigma X)^2}{n}}{n - 1}} \qquad [3\text{–}12]$$

      3. The formula for the standard deviation of grouped data is:

      $$s = \sqrt{\frac{\Sigma fM^2 - \frac{(\Sigma fX)^2}{n}}{n - 1}} \qquad [3\text{–}17]$$

Describing Data: Numerical Measures

  III. We interpret the standard deviation using two measures.
   A. Chebyshev's theorem states that regardless of the shape of the distribution, at least $1 - 1/k^2$ of the observations will be within k standard deviations of the mean, where k is greater than 1.
   B. The Empirical Rule states that for a bell-shaped distribution about 68 percent of the values will be within one standard deviation of the mean, 95 percent within two, and virtually all within three.

  IV. The coefficient of variation is a measure of relative dispersion.
   A. The formula for the coefficient of variation is:
   $$CV = \frac{s}{X}(100) \qquad [3-13]$$
   B. It reports the variation relative to the mean.
   C. It is useful for comparing distributions with different units.

  V. The coefficient of skewness measures the symmetry of a distribution.
   A. In a positively skewed set of data the long tail is to the right.
   B. In a negatively skewed distribution the long tail is to the left.

  VI. Measures of location also describe the spread in a set of observations.
   A. A quartile divides a set of observations into four equal parts.
    1. Twenty-five percent of the observations are less than the first quartile, 50 percent are less than the second quartile (the median), and 75 percent are less than the third quartile.
    2. The interquartile range is the difference between the third and the first quartile.
   B. Deciles divide a set of observations into 10 equal parts.
   C. Percentiles divide a set of observations into 100 equal parts.
   D. A box plot is a graphic display of a set of data.
    1. It is drawn enclosing the first and third quartiles.
     a. A line through the inside of the box shows the median.
     b. Dotted line segments from the third quartile to the largest value and from the first quartile to the smallest value show the range of the largest 25 percent of the observations and the smallest 25 percent.
    2. A box plot is based on five statistics: the largest and smallest observation, the first and third quartiles, and the median.

Pronunciation Key

SYMBOL	MEANING	PRONUNCIATION
μ	Population mean	mu
Σ	Operation of adding	sigma
ΣX	Adding a group of values	sigma X
$\bar{X}$	Sample mean	X bar
$\bar{X}_w$	Weighted mean	X bar sub w
GM	Geometric mean	G M
ΣfX	Adding the product of the frequencies and the class midpoints	sigma f X
σ^2	Population variance	sigma squared
σ	Population standard deviation	sigma
ΣfX^2	Sum of the product of the class midpoints squared and the class frequency	sigma f X squared
L_p	Location of percentile	L sub p
Q_1	First quartile	Q sub 1
Q_3	Third quartile	Q sub 3

Chapter 3

Chapter Exercises

73. The accounting firm of Crawford and Associates has five senior partners. Yesterday the senior partners saw six, four, three, seven, and five clients, respectively.
   a. Compute the mean number and median number of clients seen by a partner.
   b. Is the mean a sample mean or a population mean?
   c. Verify that $\Sigma(X - \mu) = 0$.

74. Owens Orchards sells apples in a large bag by weight. A sample of seven bags contained the following numbers of apples: 23, 19, 26, 17, 21, 24, 22.
   a. Compute the mean number and median number of apples in a bag.
   b. Verify that $\Sigma(X - \bar{X}) = 0$.

75. A sample of households that subscribe to a local phone company revealed the following numbers of calls received last week. Determine the mean and the median number of calls received.

52	43	30	38	30	42	12	46	39	37
34	46	32	18	41	5				

76. The Citizens Banking Company is studying the number of times the ATM, located in a Loblaws Supermarket at the foot of Market Street, is used per day. Following are the numbers of times the machine was used over each of the last 30 days. Determine the mean number of times the machine was used per day.

83	64	84	76	84	54	75	59	70	61
63	80	84	73	68	52	65	90	52	77
95	36	78	61	59	84	95	47	87	60

77. Listed below is the number of lampshades produced during the last 50 days at the Superior Lampshade Company. Compute the mean.

348	371	360	369	376	397	368	361	374
410	374	377	335	356	322	344	399	362
384	365	380	349	358	343	432	376	347
385	399	400	359	329	370	398	352	396
366	392	375	379	389	390	386	341	351
354	395	338	390	333				

78. Trudy Green works for the True-Green Lawn Company. Her job is to solicit lawn-care business via the telephone. Listed below are the number of appointments she made in each of the last 25 hours of calling. What is the arithmetic mean number of appointments she made per hour? What is the median number of appointments per hour? Write a brief report summarizing the findings.

9	5	2	6	5	6	4	4	7	2	3	6	3
4	4	7	8	4	4	5	5	4	8	3	3	

79. The Split-A-Rail Fence Company sells three types of fence to homeowners. Grade A costs $5.00/m to install, Grade B costs $6.50/m, and Grade C, the premium quality, costs $8.00/m. Yesterday, Split-A-Rail installed 270 m of Grade A, 300 m of Grade B, and 100 m of Grade C. What was the mean cost per metre of fence installed?

80. Rolland Poust is a business student. Last semester he took courses in statistics and accounting, 3 hours each, and earned an A in both. He earned a B in a five-hour history course and a B in a two-hour history of jazz course. In addition, he took a one-hour course dealing with the rules of basketball so he could get his license to officiate high school basketball games. He got an A in this course. What was his GPA for the semester? Assume that he receives 4 points for an A, 3 for a B, and so on. What measure of central tendency did you just calculate?

81. The uncertainty in the stock market led Sam to diversify his investments. However, he still felt that stock options would earn the most, and so he left the bulk of his funds in stock options. The table

Describing Data: Numerical Measures

below lists Sam's earnings from investments last year. What is the average rate of return on his investments?

Investment Type	Performance (%)	Amount Invested ($)
Mutual Funds	4.5	15 300
GICs	3.0	10 400
Stock Options	10.2	150 600

82. Listed below are the commuting distances, in kilometres, of students attending college for their first year.

| 5.2 | 6.3 | 7.5 | 4.3 | 6.8 | 4.6 | 4.6 | 8.2 | 7.8 | 9.4 | 9.3 | 7.4 | 5.3 | 5.3 | 5.4 |

    a. What is the arithmetic mean distance traveled?
    b. What is the median distance traveled?
    c. What is the modal distance traveled?

83. According to Statistics Canada, the average earnings for men in Canada have increased from $34 700 in 1993 to $38 900 in 2002, and for women, from $22 300 to $25 300 during the same time period. What is the geometric mean annual rate of increase for men and women? *(Adapted from CANSIM Table No 202-0102).*

84. Given the following quiz grades, calculate the mean, variance and standard deviation. Consider this sample data.

Grade on Quiz	3	4	5	7	8	10
Number of Students	4	4	6	20	6	1

85. The Carter Construction Company pays its summer students an hourly rate of $13.00 for general labour, $15.50 for landscaping, or $18.00 for road work. Forty students were hired for the summer, 20 for general labour, 12 for landscaping, and 8 for road work. What is the mean hourly rate paid to the 40 summer students? Calculate the variance and standard deviation.

86. A recent article suggested that if you earn $25 000 a year today and the inflation rate continues at 3 percent per year, you'll need to make $33 598 in 10 years to have the same buying power. You would need to make $44 771 if the inflation rate jumped to 6 percent. Confirm that these statements are accurate by finding the geometric mean rate of increase.

87. The ages of a sample of Canadian tourists flying from Toronto to Hong Kong were: 32, 21, 60, 47, 54, 17, 72, 55, 33, and 41.
    a. Compute the range.
    b. Compute the mean deviation.
    c. Compute the standard deviation.
    d. Using Chebyshev's Theorem, at least what percent of the observations must be within two standard deviations of the mean? Verify.
    e. Using the Empirical Rule, about 95% of the values would occur between what values? Verify.

88. The masses (in kilograms) of a sample of five boxes being sent by UPS are: 12, 6, 7, 3, and 10.
    a. Compute the range.
    b. Compute the mean deviation.
    c. Compute the standard deviation.
    d. Using Chebyshev's Theorem, at least what percent of the observations must be within 2.5 standard deviations of the mean? Verify.
    e. Using the Empirical Rule, about 68% of the values would occur between what values? Verify.

89. A library has seven branches in its system. The numbers of volumes (in thousands) held in the branches are 83, 510, 33, 256, 401, 47, and 23.
    a. Is this a sample or a population?
    b. Compute the standard deviation.
    c. Compute the coefficient of variation. Interpret.

90. Health issues are a concern of managers, especially as they evaluate the cost of medical insurance. A recent survey of 150 executives at Elvers Industries, a large insurance and financial firm, reported the number of kilograms by which the executives were overweight. Compute the range and the standard deviation.

Amount Spent ($)	Frequency
0 to under 6	14
6 to under 12	42
12 to under 18	58
18 to under 24	28
24 to under 30	8

91. A major airline wanted some information on those enrolled in their "frequent flyer" program. A sample of 48 members resulted in the following distance flown last year, in thousands of kilometres, by each participant. Develop a box plot of the data and comment on the information.

22	29	32	38	39	41	42	43	43	43	44	44
45	45	46	46	46	47	50	51	52	54	54	55
56	57	58	59	60	61	61	63	63	64	64	67
69	70	70	70	71	71	72	73	74	76	78	88

92. The National Muffler Company claims they will change your muffler in less than 30 minutes. An investigative reporter for WTOL Channel 11 monitored 30 consecutive muffler changes at the National outlet on Liberty Street. The number of minutes to perform changes is reported below.

44	12	22	31	26	22	30	26	18	28	12
40	17	13	14	17	25	29	15	30	10	28
16	33	24	20	29	34	23	13			

   a. Develop a box plot for the time to change a muffler.
   b. Does the distribution show any outliers?
   c. Summarize your findings in a brief report.

93. The Walter Gogel Company is an industrial supplier of fasteners, tools, and springs. The amounts of their invoices vary widely, from less than $20.00 to over $400.00. During the month of January they sent out 80 invoices. Here is a box plot of these invoices. Write a brief report summarizing the amounts of their invoices. Be sure to include information on the values of the first and third quartile, the median, and whether there is any skewness. If there are any outliers, approximate the value of these invoices.

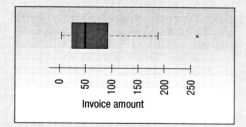

94. The following box plot shows the number of daily newspapers published. Summarize the findings. Be sure to include information on the values of the first and third quartiles, the median, and whether there is any skewness. If there are any outliers, estimate their value.

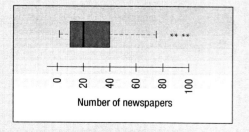

Describing Data: Numerical Measures

95. The following data are the estimated market values (in millions of dollars) of 50 companies in the auto parts business.

26.8	8.6	6.5	30.6	15.4	18.0	7.6	21.5	11.0	10.2
28.3	15.5	31.4	23.4	4.3	20.2	33.5	7.9	11.2	1.0
11.7	18.5	6.8	22.3	12.9	29.8	1.3	14.1	29.7	18.7
6.7	31.4	30.4	20.6	5.2	37.8	13.4	18.3	27.1	32.7
6.1	0.9	9.6	35.0	17.1	1.9	1.2	16.6	31.1	16.1

   a. Determine the mean and the median of the market values.
   b. Determine the standard deviation of the market values.
   c. Using the Empirical Rule, about 95 percent of the values would occur between what values?
   d. Determine the coefficient of variation.
   e. Determine the coefficient of skewness.
   f. Estimate the values of Q_1 and Q_3. Draw a box plot.
   g. Summarize the results.

96. Listed below are 20 of the best performing Canadian Equity mutual funds, their assets in millions of dollars, their three-year rates of return, and their year-to-date (Feb 28, 2005) rates of return.* Assume the data are a sample.

Fund	Assets ($ millions)	3-yr (%)	YTD (%)
Sceptre Equity Growth	346.8	33.0	9.1
Northwest Specialty Equity	236.5	32.6	12.1
Sprott Canadian Equity	975.4	28.9	5.3
Acuity All Cap 30 Canadian Equity	94.3	24.4	0.8
Acuity Pooled Canadian Equity	39.2	22.6	6.1
R Small Cap Canadian Equity	64.4	20.8	6.2
BluMont Hirsch Performance	83.6	19.8	6.9
Dynamic Power Canadian Growth	1291.9	19.5	12.9
Acuity Canadian Equity	51.1	19.1	5.8
Dynamic Canadian Value Class	29.6	19.0	10.5
Dynamic Power Canadian Growth Class	17.0	18.8	12.1
Halcyon Hirsch Opportunistic Canadian	19.4	18.6	6.6
ABC Fundamental-Value	517.0	18.2	8.1
Dynamic Focus + Small Business	25.8	17.3	11.2
Acuity Social Values Canadian Equity	27.5	16.8	6.9
Synergy Extreme Canadian Equity	60.7	16.7	3.9
RBC O'Shaughnessy Canadian Equity	1357.0	16.0	5.1
Beutel Goodman Small Cap	162.0	14.9	5.9
HSBC Small Cap Growth	49.7	14.8	6.1
Mac Growth	436.1	14.6	3.5

Source: Morningstar Canada * Performance data is as of Feb. 28, 2005.

   a. Compute the mean, median, and standard deviation for each of the variables. Compare the standard deviations for the year-to-date and three-year rates of return. Comment on your findings.
   b. Compute the coefficient of variation for each of the above variables. Comment on the relative variation of the three variables.
   c. Compute the coefficient of skewness for each of the above variables. Comment on the skewness of the three variables.
   d. Compute the first and third quartiles for the year-to-date and three-year rates of return.
   e. Draw a box plot for the year-to-date and the three-year rates of return. Comment on the results. Are there any outliers?

97. The Apollo space program lasted from 1967 until 1972 and included 13 missions. The missions lasted from as little as 7 hours to as long as 301 hours. The duration of each flight is listed below.

9	195	241	301	216	260	7	244	192	147
10	295	142							

Chapter 3

a. Find the mean, median, and standard deviation of the duration for the Apollo flights. Treat as sample data.
b. Compute the coefficient of variation and the coefficient of skewness. Comment on your findings.
c. Find the 45th and 82nd percentiles.
d. Draw a box plot and comment on your findings.

98. A recent report in *Woman's World* magazine suggested that the typical family of four with an intermediate budget spends about $96 per week on food. The following frequency distribution was included in the report. Compute the mean and the standard deviation.

Amount Spent ($)	Frequency
80 to under 85	6
85 to under 90	12
90 to under 95	23
95 to under 100	35
100 to under 105	24
105 to under 110	10

99. Bidwell Electronics, Inc., recently surveyed a sample of employees to determine how far they lived from corporate headquarters. The results are shown below. Compute the mean and the standard deviation.

Distance (km)	Frequency
0 to under 5	4
5 to under 10	15
10 to under 15	27
15 to under 20	18
20 to under 25	6

100. A survey showed that in a class of 30 students, nine had purchased their own computers. The cost of the computers, in dollars, is listed below.

| 2235 | 2150 | 1850 | 1500 | 2025 | 5750 | 2800 | 2750 | 3300 |

a. Calculate the mean and median cost of the computers.
b. Draw a box plot and comment on your findings.
c. Would you use the mean or median as a measure of centre of your data? Explain.

Data Set Exercises

101. Refer to the CREA data on the CD-ROM, which reports information on average house prices nationally and in a selection of cities across Canada for January and March, 2004 and 2005.
 a. Select the cities only for the variable Jan-04.
  1. Find the mean, median, and the standard deviation.
  2. Determine the coefficient of skewness. Is the distribution positively or negatively skewed?
  3. Develop a box plot. Are there any outliers? Estimate the first and third quartiles.
  4. Summarize the results.
 b. Select the cities only for the variable Jan-05.
  1. Find the mean, median, and the standard deviation.
  2. Determine the coefficient of skewness. Is the distribution positively or negatively skewed?
  3. Develop a box plot. Are there any outliers? Estimate the first and third quartiles.
  4. Summarize the results.

102. Refer to the Real Estate data, Regina & Surrounding Area, on the CD-ROM, which reports information on listed homes and townhomes, March 2005.
 a. Select the variable size (sq ft).
  1. Find the mean, median, and the standard deviation.
  2. Determine the coefficient of skewness. Is the distribution positively or negatively skewed?

Describing Data: Numerical Measures

   3. Develop a box plot. Are there any outliers? Estimate the first and third quartiles.
   4. Summarize the results.
  b. Select the variable no. of bedrooms.
   1. Find the mean, median, and the standard deviation.
   2. Determine the coefficient of skewness. Is the distribution positively or negatively skewed?
   3. Develop a box plot. Are there any outliers? Estimate the first and third quartiles.
   4. Summarize the results.

103. Refer to the International data, which reports demographic and economic information on 46 countries.
  a. Select the variable Life Expectancy.
   1. Find the mean, median, and the standard deviation.
   2. Determine the coefficient of skewness. Is the distribution positively or negatively skewed?
   3. Develop a box plot. Are there any outliers? Estimate the first and third quartiles.
   4. Summarize the results.
  b. Select the variable GDP/cap.
   1. Find the mean, median, and the standard deviation.
   2. Determine the coefficient of skewness. Is the distribution positively or negatively skewed?
   3. Develop a box plot. Are there any outliers? Estimate the first and third quartiles.
   4. Summarize the results.
  c. Select the variable Labour Force.
   1. Find the mean, median, and the standard deviation.
   2. Determine the coefficient of skewness. Is the distribution positively or negatively skewed?
   3. Develop a box plot. Are there any outliers? Estimate the first and third quartiles.
   4. Summarize the results.

Case

Continue with the Whitner Pontiac data and case from Chapter 2. Rob would like you to further develop tables and charts that he could review monthly and would like you to report where the selling prices tend to cluster, where the variation is in the selling prices, and to note any trends. The data is on the CD-ROM, Data Files, Whitner-2005.

Additional exercises that require you to access information at related Internet sites are available on the CD-ROM included with this text.

Chapter 3 Answers to Self-Reviews

3–1 1. (a) $\bar{X} = \dfrac{\$267\,100}{4} = \$66\,775$

  (b) Statistic, because it is a sample value.

  (c) $66 775. The sample mean is our best estimate of the population mean.

  2. (a) $\mu = \dfrac{498}{6} = 83$

  (b) Parameter, because it was computed using all the population values.

3–2 (a) $237, found by:

$$\dfrac{(95 \times \$400) + (126 \times \$200) + (79 \times \$100)}{95 + 126 + 79} = \$237.00$$

  (b) The profit per suit is $12, found by $237 − $200 cost − $25 commission. The total profit for the 300 suits is $3600, found by 300 × $12.

3–3 1. (a) $284.50
  (b) 3, 3
  2. (a) 7, found by (6 + 8)/2 = 7
  (b) 3, 3
  (c) 0

3–4 (a)

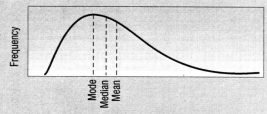

Weekly sales ($ thousands)

  (b) Positively skewed, because the mean is the largest average and the mode is the smallest.

3–5 1. (a) About 8.39 percent, found by $\sqrt[4]{4951.75464}$
  (b) About 10.095 percent
  (c) Greater than, because 10.095 > 8.39

  2. 8.63 percent, found by $\sqrt[20]{\dfrac{120\,520}{23\,000}} - 1 = 1.0863 - 1$

3–6 (a) 22, found by 112 − 90

  (b) $\bar{X} = \dfrac{824}{8} = 103$

  (c)

X	$|X - \bar{X}|$	Absolute Deviation
95	$|-8|$	8
103	$|0|$	0
105	$|+2|$	2
110	$|+7|$	7
104	$|+1|$	1
105	$|+2|$	2
112	$|+9|$	9
90	$|-13|$	13
		Total 42

$MD = \dfrac{42}{8} = 5.25\,\text{kg}$

3–7 (a) $\mu = \dfrac{\$16\,900}{5} = \$3380$

  (b) $\sigma^2 = \dfrac{(3536 - 3380)^2 + \cdots + (3622 - 3380)^2}{5}$

  $= \dfrac{197\,454}{5} = 39\,490.8$

  (c) $\sigma = \sqrt{39\,490.8} = 198.72$

  (d) There is more variation in the second office because the standard deviation is larger. The mean is also larger in the second office.

3–8 2.33, found by:

$\bar{X} = \dfrac{\Sigma X}{n} = \dfrac{28}{7} = 4$

X	X − $\bar{X}$	$(X - \bar{X})^2$	X^2
4	0	0	16
2	−2	4	4
5	1	1	25
4	0	0	16
5	1	1	25
2	−2	4	4
6	2	4	36
28	0	14	126

$s^2 = \dfrac{\Sigma(X - \bar{X})^2}{n - 1}$ or $s^2 = \dfrac{\Sigma X^2 - \dfrac{(\Sigma X)^2}{n}}{n - 1}$

$= \dfrac{14}{7 - 1}$ $= \dfrac{126 - \dfrac{(28)^2}{7}}{7 - 1}$

$= 2.33$ $= \dfrac{126 - 112}{6}$

$= 2.33$

$s = \sqrt{2.33} = 1.53$

3–9 (a) $k = \dfrac{1.35 - 1.2}{.1} = 1.5$

$1 - \dfrac{1}{(1.5)^2} = 1 - .44 = .56$

$= 56\%$

  (b) 1.1 to 1.3

  (c) 1.0 to 1.4

3–10 CV for mechanical is 5 percent, found by (10/200)(100). For finger dexterity, CV is 20 percent, found by (6/30)(100). Thus, relative dispersion in finger dexterity scores is greater than relative dispersion in mechanical, because 20 percent > 5 percent.

Describing Data: Numerical Measures

3–11 (a) $\bar{X} = \dfrac{407}{5} = 81.4$, Median $= 84$

$$s = \sqrt{\dfrac{34\,053 - \dfrac{(407)^2}{5}}{5-1}} = 15.19$$

(b) $sk = \dfrac{3(81.4 - 84.0)}{15.19} = -0.51$

(c) The distribution is somewhat negatively skewed.

3–12 (a) 500

(b) $Q_1 = 495.5$, $Q_2 = 502.5$

3–13 The smallest value is 10 and the largest 85; the first quartile is 25 and the third 60. About 50 percent of the values are between 25 and 60. The median value is 40. The distribution is somewhat positively skewed.

3–14 a. Frequency distribution.

b.

f	M	fM	fM²
1	4	4	16
4	8	32	256
10	12	120	1440
3	16	48	768
2	20	40	800
20		244	3280

$$\bar{X} = \dfrac{\Sigma fM}{M} = \dfrac{\$244}{20} = \$12.20$$

c. $s = \sqrt{\dfrac{3280 - \dfrac{(244)^2}{20}}{20-1}} = \$3.99$

Chapter 6

The Normal Probability Distribution

LEARNING OBJECTIVES

When you have completed this chapter, you will be able to:

1. List the characteristics of the normal probability distribution.

2. Define and calculate z values.

3. Determine the probability that an observation is between two points on a normal distribution using the standard normal distribution.

4. Determine the probability that an observation is above (or below) a point on a normal distribution using the standard normal distribution.

5. Compare two or more observations that are on different probability distributions.

6. Use the normal probability distribution to approximate the binomial probability distribution.

Introduction

Chapter 5 began our study of probability distributions. We considered three discrete probability distributions: the binomial, hypergeometric, and the Poisson. These distributions are based on discrete random variables, which can assume only clearly separated values. For example, the number of correct answers on a 10-question examination can only be 0, 1, 2, 3, . . . , 10. There cannot be a negative number of correct answers, such as −7, nor can there be 7¼ or 15 correct answers.

In this example, only certain outcomes are possible and these outcomes are represented by clearly separated values. In addition, the result is usually found by counting the number of successes or the number of questions answered correctly.

In this chapter, we continue our study of probability distributions by examining a very important *continuous* probability distribution, the **normal probability distribution**. A continuous probability distribution usually results from measuring something, such as the distance from the residence to the classroom, the weight of an individual, or the amount of bonuses earned by CFOs. Suppose we select five students and find the distance, in kilometres, that they travel to attend class as 12.2, 8.9, 6.7, 3.6, and 14.6. When examining a continuous distribution we are usually interested in information such as the percent of students who travel less than 10 km or the percent of students who travel more than 8 km, or perhaps the percent of observations that occur within a certain range. Then, the number of kilometres traveled could have an infinite number of values within a particular range. So you think of the probability a variable will have a value within a specified range rather than the probability for a specific value.

The normal probability distribution describes the likelihood that a continuous random variable with an infinite number of possible values lies within a specified range. For example, assume the life of an Energizer C size battery follows a normal distribution with a mean of 45 hours and a standard deviation of 10 hours when used in a particular toy. We can determine the likelihood the battery will last more than 50 hours, between 35 and 62 hours, or less than 39 hours.

The Normal Probability Distribution

The Family of Normal Probability Distributions

The normal probability distribution and its accompanying normal curve have the following major characteristics:

1. It is **bell-shaped** and has a single peak in the centre of the distribution. The arithmetic mean, median, and mode are equal and located in the centre of the distribution. Thus, half the area under the curve is to the right of this centre point, and the other half is to the left of it.
2. It is **symmetrical** about the mean. If we cut the normal curve vertically at this central value, the two halves will be mirror images.
3. It falls off smoothly in either direction from the central value. It is **asymptotic,** meaning that the curve gets closer and closer to the X-axis but never actually touches it. That is, the "tails" of the curve extend indefinitely in both directions.
4. With a continuous probability distribution, areas below the curve define probabilities. The total area under the normal curve is 100%.

These characteristics are shown graphically in Chart 6–1.

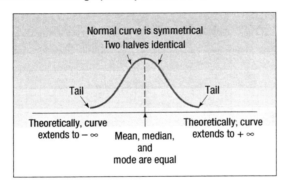

CHART 6–1 Characteristics of a Normal Distribution

For example, in Chart 6–2 the probability distributions of length of employee service in three different plants can be compared. There is not just one normal probability distribution, but rather a "family" of them. In the Camden plant, the mean is 20 years and the standard deviation is 3.1 years. There is another normal probability distribution for the length of service in the Dunkirk plant, where $\mu = 20$ years and $\sigma = 3.9$ years. In the Elmira plant, $\mu = 20$ years and $\sigma = 5.0$. Note that the means are the same but the standard deviations are different.

Chart 6–3 shows the distribution of box masses of three different cereals. The masses are normally distributed with different means but identical standard deviations.

Equal means, unequal standard deviations

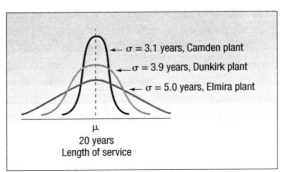

CHART 6–2 Normal Probability Distributions with Equal Means but Different Standard Deviations

Unequal means, equal standard deviations

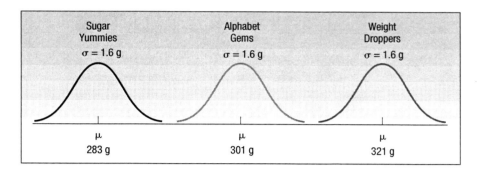

CHART 6–3 Normal Probability Distributions Having Different Means but Equal Standard Deviations

Finally, Chart 6–4 shows three normal distributions having different means and standard deviations. They show the distribution of tensile strengths, measured in kg/m², for three types of cables.

Unequal means, unequal standard deviations

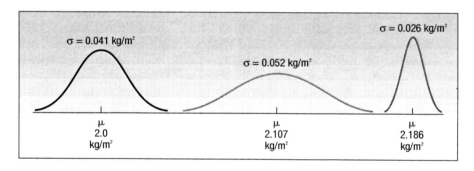

CHART 6–4 Normal Probability Distributions with Different Means and Standard Deviations

In Chapter 5, recall that discrete probability distributions show the specific likelihood a discrete value will occur. For example, the binomial distribution is used to calculate the probability that none of five flights arriving at the Calgary Airport would be late.

As stated previously, with a continuous probability distribution, areas below the curve define probabilities. The total area under the normal curve is 1.0. This accounts for all possible outcomes. Since a normal probability distribution is symmetric, the area under the curve to the left of the mean is 0.5, and the area under the curve to the right of the mean is 0.5. Apply this to the distribution of Sugar Yummies in Chart 6–3. It is normally distributed with a mean of 283 g. Therefore, the probability of filling a box with more than 283 g is 0.5 and the probability of filling a box with less than 283 g is 0.5. We can also determine the probability that the mass of a box is between 280 g and 286 g. However, to determine this probability we need to know about the standard normal probability distribution.

The Standard Normal Distribution

The number of normal distributions is unlimited, each having a different mean (μ), standard deviation (σ), or both. While it is possible to provide probability tables for discrete distributions such as the binomial and the Poisson, providing tables for the infinite number of normal distributions is impossible. Fortunately, one member of the family can be used to

The Normal Probability Distribution

determine the probabilities for all normal distributions. It is called the **standard normal distribution**, and it is unique because it has a mean of 0 and a standard deviation of 1.

Any normal distribution can be converted into a *standard normal distribution* by subtracting the mean from each observation and dividing this difference by the standard deviation. The results are called **z values**. They are also referred to as **z scores**, the **z statistic**, the **standard normal deviate**, or just the **normal deviate**.

> **z VALUE** The signed distance between a selected value, designated X, and the mean, μ, divided by the standard deviation, σ.

So, a z value is the distance from the mean, measured in units of the standard deviation.
In terms of a formula:

STANDARD NORMAL VALUE $$z = \frac{X - \mu}{\sigma}$$ [6–1]

where:

- X is the value of any particular observation or measurement.
- μ is the mean of the distribution.
- σ is the standard deviation of the distribution.

As noted in the above definition, a z value expresses the distance or difference between a particular value of X and the arithmetic mean in units of the standard deviation. Once the normally distributed observations are standardized, the z values are normally distributed with a mean of 0 and a standard deviation of 1. The table in Appendix D (also on the inside back cover) lists the probabilities for the standard normal probability distribution.

To explain, suppose we wish to compute the probability that boxes of Sugar Yummies have a mass between 283 g and 285.4 g. From Chart 6–3, we know that the box mass of Sugar Yummies follows the normal distribution with a mean of 283 g and a standard deviation of 1.6 g. We want to know the probability or area under the curve between the mean, 283 g, and 285.4 g. We can also express this problem using probability notation, similar to the style used in the previous chapter: $P(283 < \text{mass} < 285.4)$. To find the probability, it is necessary to convert both 283 g and 285.4 g to z values using formula (6–1). The z value corresponding to 283 is 0, found by $(238 - 283)/1.6$. The z value corresponding to 285.4 is 1.50 found by $(285.4 - 283)/1.6$. Next we go to the table in Appendix D. A portion of the table is repeated as Table 6–1. Go down the column of the table headed by the letter z to 1.5. Then move horizontally to the right and read the probability under the column headed 0.00. It is 0.4332. This means the area under the curve between 0.00 and 1.50 is 0.4332. This is the probability that a randomly selected box of Sugar Yummies will have a mass between 283 g and 285.4 g.

TABLE 6–1 Areas under the Normal Curve

z	0.00	0.01	0.02	0.03	0.04	0.05	...
1.3	0.4032	0.4049	0.4066	0.4082	0.4099	0.4115	
1.4	0.4192	0.4207	0.4222	0.4236	0.4251	0.4265	
1.5	0.4332	0.4345	0.4357	0.4370	0.4382	0.4394	
1.6	0.4452	0.4463	0.4474	0.4484	0.4495	0.4505	
1.7	0.4554	0.4564	0.4573	0.4582	0.4591	0.4599	
1.8	0.4641	0.4649	0.4656	0.4664	0.4671	0.4678	
1.9	0.4713	0.4719	0.4726	0.4732	0.4738	0.4744	
⋮							

170 Chapter 6

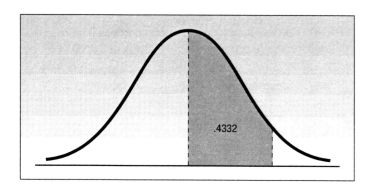

Now we will compute the z value given the population mean, μ, the population standard deviation, σ, and a selected X.

EXAMPLE

The weekly incomes of shift supervisors in the glass industry are normally distributed with a mean of $1000 and a standard deviation of $100. What is the z value for the income X of a supervisor who earns $1100 per week? For a supervisor who earns $900 per week?

Solution

Using formula (6–1), the z values for the two X values ($1100 and $900) are:

For X = $1100:
$$z = \frac{X - \mu}{\sigma}$$
$$= \frac{\$1100 - \$1000}{\$100}$$
$$= 1.00$$

For X = $900:
$$z = \frac{X - \mu}{\sigma}$$
$$= \frac{\$900 - \$1000}{\$100}$$
$$= -1.00$$

The z of 1.00 indicates that a weekly income of $1100 is one standard deviation above the mean, and a z of −1.00 shows that a $900 income is one standard deviation below the mean. Note that both incomes ($1100 and $900) are the same distance ($100) from the mean.

Self-Review 6–1

Using the information in the preceding example ($\mu = \$1000$, $\sigma = \$100$), convert:
(a) The weekly income of $1225 to a z value.
(b) The weekly income of $775 to a z value.

Statistics in Action

An individual's skills depend on a combination of many hereditary and environmental

The Empirical Rule

Before examining further applications of the standard normal probability distribution, we will consider three areas under the normal curve that will be used extensively in the following chapters. These facts were called the Empirical Rule in Chapter 3.

1. About 68 percent of the area under the normal curve is within one standard deviation of the mean. This can be written as $\mu \pm 1\sigma$. The actual percentage is 68.26%.
2. About 95 percent of the area under the normal curve is within two standard deviations of the mean, written $\mu \pm 2\sigma$. The actual percentage is 95.44%.
3. Practically all of the area under the normal curve is within three standard deviations of the mean, written $\mu \pm 3\sigma$. The actual percentage is 99.7%.

The Normal Probability Distribution

factors, each having about the same amount of weight or influence on the skills. Thus, much like a binomial distribution with a large number of trials, many skills and attributes follow the normal distribution. For example, scores on the Scholastic Aptitude Test (SAT) are normally distributed with a mean of 1000 and a standard deviation of 140.

This information is summarized in the following graph.

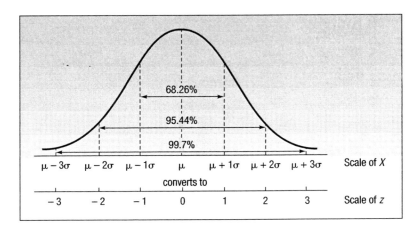

Transforming measurements to standard normal deviates changes the scale. The conversions are also shown in the graph. For example, $\mu + 1\sigma$ is converted to a z value of $+1.00$. Likewise, $\mu - 2\sigma$ is transformed to a z value of -2.00. Note that the centre of the z distribution is zero, indicating no deviation from the mean, μ.

EXAMPLE

As part of their quality assurance program, the Autolite Battery Company conducts tests on battery life. For a particular D cell alkaline battery, the mean life is 19 hours. The useful life of the battery follows a normal distribution with a standard deviation of 1.2 hours. Answer the following questions.

1. $\mu \pm 1\sigma$ of the batteries failed between what two values?
2. $\mu \pm 2\sigma$ of the batteries failed between what two values?
3. $\mu \pm 3\sigma$ of the batteries failed between what two values?

Solution

We can use the results of the Empirical Rule to answer these questions.

1. $\mu \pm 1\sigma$ of the batteries will fail between 17.8 and 20.2 hours, found by $19.0 \pm 1(1.2)$ hours.
2. $\mu \pm 2\sigma$ of the batteries will fail between 16.6 and 21.4 hours, found by $19.0 \pm 2(1.2)$ hours.
3. $\mu \pm 3\sigma$ failed between 15.4 and 22.6 hours, found by $19.0 \pm 3(1.2)$ hours.

This information is summarized on the following chart.

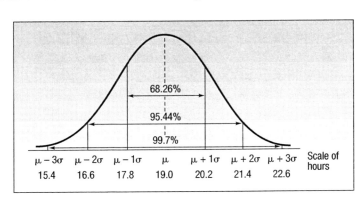

172   Chapter 6

Self-Review 6–2

The distribution of the annual incomes of a group of middle-management employees at Compton Plastics approximates a normal distribution with a mean of $47 200 and a standard deviation of $800.

(a) $\mu \pm 1\sigma$ of the incomes lie between what two amounts?
(b) $\mu \pm 2\sigma$ of the incomes lie between what two amounts?
(c) $\mu \pm 3\sigma$ of the incomes lie between what two amounts?
(d) What are the median and the modal incomes?
(e) Is the distribution of incomes symmetrical?

Exercises

1. The mean of a normal probability distribution is 500; the standard deviation is 10.
   a. $\mu \pm 1\sigma$ of the observations lie between what two values?
   b. $\mu \pm 2\sigma$ of the observations lie between what two values?
   c. $\mu \pm 3\sigma$ of the observations lie between what two values?
2. The mean of a normal probability distribution is 60; the standard deviation is 5.
   a. What percent of the observations lie between 55 and 65?
   b. What percent of the observations lie between 50 and 70?
   c. What percent of the observations lie between 45 and 75?
3. The Kamp family has twins, Rob and Rachel. Both Rob and Rachel graduated from college 2 years ago, and each is now earning $50 000 per year. Rachel works in the retail industry, where the mean salary for executives with less than 5 years' experience is $35 000 with a standard deviation of $8000. Rob is an engineer. The mean salary for engineers with less than 5 years' experience is $60 000 with a standard deviation of $5000. Compute the z values for both Rob and Rachel and comment on your findings.
4. A recent article in the Sherkston Beach *Sun Times* reported that the mean labour cost to repair a colour TV is $90 with a standard deviation of $22. Dawson's Electronic Repairs and Services completed repairs on two sets this morning. The labour cost for the first was $75 and it was $100 for the second. Compute z values for each and comment on your findings.

Finding Areas under the Normal Curve

The first application of the standard normal distribution involves finding the area in a normal distribution between the mean and a selected value, which we identify as X. The following example will illustrate the details.

EXAMPLE

Recall in an earlier example we reported that the mean weekly income of a shift supervisor in the glass industry is normally distributed with a mean of $1000 and a standard deviation of $100. That is, $\mu = \$1000$ and $\sigma = \$100$. What is the likelihood of selecting a supervisor whose weekly income is between $1000 and $1100? We write this question in probability notation as: P($1000 < weekly income < $1100).

Solution

We have already converted $1100 to a z value of 1.00 using formula (6–1). To repeat:

$$z = \frac{X - \mu}{\sigma} = \frac{\$1100 - \$1000}{\$100} = 1.00$$

The probability associated with a z of 1.00 is available in Appendix D. A portion of Appendix D follows. To locate the probability, go down the left column to 1.0, and then move horizontally to the column headed .00. The value is .3413.

z	.00	.01	.02
⋮	⋮	⋮	⋮
0.7	.2580	.2611	.2642
0.8	.2881	.2910	.2939
0.9	.3159	.3186	.3212
1.0	.3413	.3438	.3461
1.1	.3643	.3665	.3686
⋮	⋮	⋮	⋮

The Normal Probability Distribution

Statistics in Action

Many processes, such as filling soft drink bottles and canning fruit, are normally distributed. Manufacturers must guard against both over- and underfilling. If they put too much in the can or bottle, they are giving away their product. If they put too little in, the customer may feel cheated and the government may question the label description. "Control Charts," with limits drawn three standard deviations above and below the mean, are routinely used to monitor this type of production process.

The area under the normal curve between $1000 and $1100 is .3413. We could also say 34.13 percent of the shift supervisors in the glass industry earn between $1000 and $1100 weekly, or the likelihood of selecting a supervisor and finding his or her income is between $1000 and $1100 is .3413.

This information is summarized in the following chart.

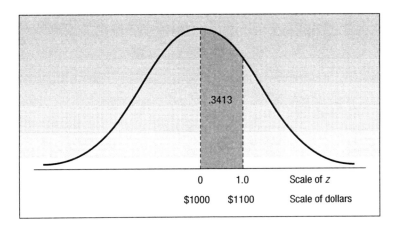

In the example just completed, we are interested in the probability between the mean and a given value. Let's change the question. Instead of wanting to know the probability of selecting a supervisor at random who earned between $1000 and $1100, suppose we wanted the probability of selecting a supervisor who earned less than $1100. In probability notation we write this statement as P(weekly income < $1100). The method of solution is the same. We find the probability of selecting a supervisor who earns between $1000, the mean, and $1100. This probability is .3413. Next, recall that half the area, or probability, is above the mean and half is below. So the probability of selecting a supervisor earning less than $1000 is .5000. Finally, we add the two probabilities, so .3413 + .5000 = .8413. About 84 percent of the supervisors in the glass industry earn less than $1100 per month. See the following diagram.

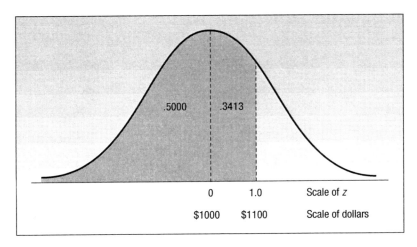

Computer software will calculate this probability.

The Excel and MegaStat commands to find the output for the weekly incomes of shift supervisors follows. Both Excel and MegaStat will calculate the area to the left (lower) and to the right (upper) of the point $1100.

174   Chapter 6

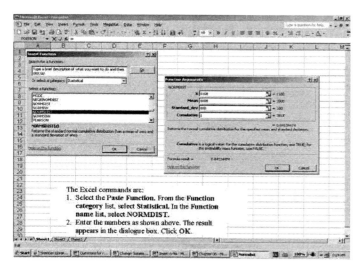

The Excel commands are:
1. Select the Paste Function. From the Function category list, select Statistical. In the Function name list, select NORMDIST.
2. Enter the numbers as shown above. The result appears in the dialogue box. Click OK.

The answer to the above problem using Excel is the area in the lower tail, .8413. Since NORMDIST is a cumulative function, it measures the total area to the left of $1100. However, if you want to find the area under the normal curve between $1000 (the mean) and $1100, we subtract the area to the left of the mean (.5). This gives us .8413 − .5 = .3413.

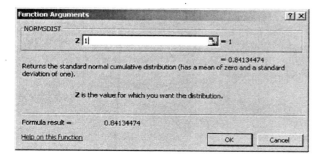

If the z-value is known, then the Excel function NORMSDIST can be used. The function gives the same result as noted in the dialogue box above.

The MegaStat commands follow.

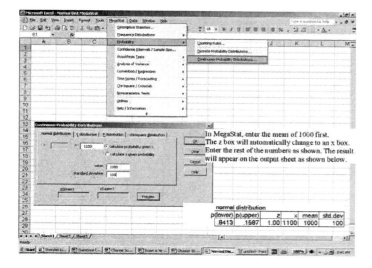

The Normal Probability Distribution

As with Excel, MegaStat will return the area in the lower tail as .8413. Since this is a cumulative function, it measures the total area to the left of $1100. To find the area under the normal curve between $1000 (the mean) and $1100, we subtract the area to the left of the mean (.5). This gives us .8413 − .5 = .3413.

EXAMPLE

Refer to the information regarding the weekly income of shift supervisors in the glass industry. The distribution of weekly incomes follows the normal distribution, with a mean of $1000 and a standard deviation of $100. What is the probability of selecting a shift supervisor in the glass industry whose income is:

1. Between $790 and $1000?
2. Less than $790?

Solution

We begin by finding the z value corresponding to a weekly income of $790. From formula (6–1):

$$z = \frac{X - \mu}{\sigma} = \frac{\$790 - \$1000}{\$100} = -2.10$$

See Appendix D. Move down the left margin to the row 2.1 and across that row to the column headed 0.00. The value is .4821. So the area under the standard normal curve corresponding to a z value of 2.10 is .4821. However, because the normal distribution is symmetric, the area between 0 and a negative z is the same as that between 0 and z. The likelihood of finding a supervisor earning between $790 and $1000 is .4821. In probability notation we write $P(\$790 < \text{weekly income} < \$1000) = .4821$.

z	0.00	0.01	0.02
⋮	⋮	⋮	⋮
2.0	.4772	.4778	.4783
2.1	.4821	.4826	.4830
2.2	.4861	.4864	.4868
2.3	.4893	.4896	.4898
⋮	⋮	⋮	⋮

The mean divides the normal curve into two identical halves. The area under the half to the left of the mean is .5000, and the area to the right is also .5000. Because the area under the curve between $790 and $1000 is .4821, the area below $790 is .0179, found by .5000 − .4821. In probability notation we write $P(\text{weekly income} < \$790) = .0179$.

This means that 48.21 percent of the supervisor have weekly incomes between $790 and $1000. Further, we can anticipate that 1.79 percent earn less than $790 per week. This information is summarized in the following diagram.

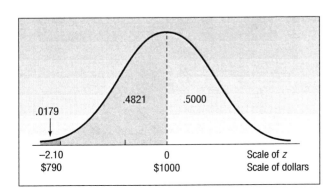

176   Chapter 6

Excel and MegaStat are cumulative functions, and will split the area under the normal curve into two areas at the point in question. So, to find the required probability it may be necessary to add or subtract areas. The MegaStat commands for the above example follow.

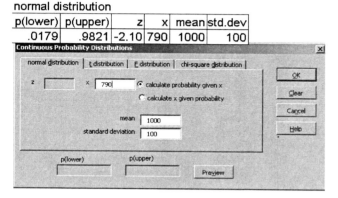

MegaStat returns .9821 in the upper tail. This is the area under the curve from $790 to the right. To find the area between $790 and $1000, we must subtract the area from $1000 to the right, which is .5. So, the result is .9821 − .5 = .4821.

Self-Review 6–3

The employees of Cartwright Manufacturing are awarded efficiency ratings. The distribution of the ratings follows a normal distribution. The mean is 400, the standard deviation 50.

(a) What is the area under the normal curve between 400 and 482? Write this area in probability notation.
(b) What is the area under the normal curve for ratings greater than 482? Write this area in probability notation.
(c) Show the details of this problem in a chart.

Exercises

5. Given the following z-values, find the areas under the normal curve to the left and right of the z-value.
   a. $z = 1.0$
   b. $z = -1.25$
   c. $z = 2.83$
   d. $z = -2.35$
6. Given the following pairs of z-values, find the area under the normal curve between each pair of z-values.
   a. $z = 1.0$ and $z = -1.0$
   b. $z = -1.25$ and $z = -2.0$
   c. $z = 2.83$ and $z = 1.75$
   d. $z = -2.35$ and $z = 1.5$
7. A normal population has a mean of 20.0 and a standard deviation of 4.0.
   a. Compute the z value associated with 25.0.
   b. What proportion of the population is between 20.0 and 25.0?
   c. What proportion of the population is less than 18.0?
8. A normal population has a mean of 12.2 and a standard deviation of 2.5.
   a. Compute the z value associated with 14.3.
   b. What proportion of the population is between 12.2 and 14.3?
   c. What proportion of the population is less than 10.0?
9. A recent study of the hourly wages of maintenance crews for major airlines showed that the mean hourly salary was $16.50, with a standard deviation of $3.50. If we select a crew member at random, what is the probability the crew member earns:
   a. Between $16.50 and $20.00 per hour?
   b. More than $20.00 per hour?
   c. Less than $15.00 per hour?
10. The mean of a normal distribution is 400 kg. The standard deviation is 10 kg.
    a. What is the area between 415 kg and the mean of 400 kg?
    b. What is the area between the mean and 395 kg?
    c. What is the probability of selecting a value at random and discovering that it has a value of less than 395 kg?

The Normal Probability Distribution

A second application of the normal distribution involves combining two areas, or probabilities. One of the areas is to the right of the mean and the other to the left.

EXAMPLE

Recall the distribution of weekly incomes of shift supervisors in the glass industry. The weekly incomes follow the normal distribution, with a mean of $1000 and a standard deviation of $100. What is the area under this normal curve between $840 and $1200?

Solution

The problem can be divided into two parts. For the area between $840 and the mean of $1000:

$$z = \frac{\$840 - \$1000}{\$100} = \frac{-\$160}{\$100} = -1.60$$

For the area between the mean of $1000 and $1200:

$$z = \frac{\$1200 - \$1000}{\$100} = \frac{-\$200}{\$100} = 2.00$$

The area under the curve for a z of −1.60 is .4452 (from Appendix D). The area under the curve for a z of 2.00 is .4772. Adding the two areas: .4452 + .4772 = .9224. Thus, the probability of selecting an income between $840 and $1200 is .9224. In probability notation we write P($840 < weekly income < $1200) = .4452 + .4772 = .9224. To summarize, 92.24 percent of the supervisors have weekly incomes between $840 and $1200. This is shown in a diagram:

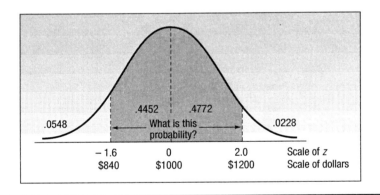

The output from MegaStat for the values of $840 and $1200 follow.
Note that the area in the two tails of the above example are calculated as .5 − .4452 = .0548 in the lower tail and .5 − .4772 = .0228 in the upper tail.

normal distribution

p(lower)	p(upper)	z	x	mean	std.dev
.0548	.9452	-1.60	840	1000	100

normal distribution

p(lower)	p(upper)	z	x	mean	std.dev
.9772	.0228	2.00	1200	1000	100

The output from MegaStat for the value of $840 is .0548 in the lower tail and .9452 in the upper tail. The output for the value of $1200 is .9772 in the lower tail and .0228 in the upper tail. To obtain the required result, take the total area and subtract the

178   Chapter 6

areas that you do not need. (This is using the complement rule). The required result is $1 - .0548 - .0228 = .9224$.

Another application of the normal distribution involves determining area between values on the *same* side of the mean.

EXAMPLE

Returning to the weekly income distribution of shift supervisors in the glass industry ($\mu = \$1000$, $\sigma = \$100$), what is the area under the normal curve between $1150 and $1250?

Solution

The situation is again separated into two parts, and formula (6–1) is used. First, we find the z value associated with a weekly salary of $1250:

$$z = \frac{\$1250 - \$1000}{\$100} = 2.50$$

Next we find the z value for a weekly salary of $1150:

$$z = \frac{\$1150 - \$1000}{\$100} = 1.50$$

From Appendix D the area associated with a z value of 2.50 is .4938. So the probability of a weekly salary between $1000 and $1250 is .4938. Similarly, the area associated with a z value of 1.50 is .4332, so the probability of a weekly salary between $1000 and $1150 is .4332. The probability of a weekly salary between $1150 and $1250 is found by subtracting the area associated with a z value of 1.50 (.4332) from that associated with a z of 2.50 (.4938). Thus, the probability of a weekly salary between $1150 and $1250 is .0606. In probability notation we write $P(\$1150 \leq \text{weekly income} \leq \$1250) = .4938 - .4332 = .0606$.

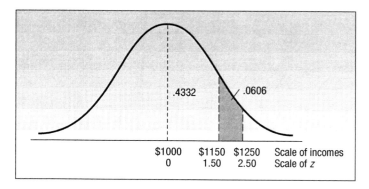

This example is similar to the previous example using MegaStat. The output from MegaStat for the value of $1150 is .9332 in the lower tail and .0668 in the upper tail. The output for the value of $1250 is .9938 in the lower tail and .0062 in the upper tail. Again we will use the complement rule and subtract the areas that we do not need. The required result is $1 - .9332 - .0062 = .0606$.

normal distribution

p(lower)	p(upper)	z	x	mean	std.dev
.9332	.0668	1.50	1150	1000	100

normal distribution

p(lower)	p(upper)	z	x	mean	std.dev
.9938	.0062	2.50	1250	1000	100

The Normal Probability Distribution

In brief, there are four situations for finding the area under the standard normal distribution.

1. To find the area between 0 and z (or −z), look up the probability directly in the table.
2. To find the area beyond z or (−z), locate the probability of z in the table and subtract that probability from .5000.
3. To find the area between two points on different sides of the mean, determine the z values and add the corresponding probabilities.
4. To find the area between two points on the same side of the mean, determine the z values and subtract the smaller probability from the larger.

EXAMPLE

The average list price of 90 homes in Regina and the surrounding area is $192 712 with a standard deviation of $86 881. What is the probability that:

1. a house is listed for more than $350 000?
2. a house is listed for between $150 000 and $300 000?
3. a house is listed for between $75 000 and $125 000?

Solution

1. We begin by finding the z-value corresponding to a list price of $350 000. From formula (6–1):

$$z = \frac{X - \mu}{\sigma} = \frac{350\,000 - 192\,712}{86\,881} = 1.81$$

See Appendix D. Move down the left margin to the row headed 1.8 and across that row to the column headed 0.01. The value is .4649. However, the required area is to the right of $350 000, so we subtract .4649 from .5. Then .5 − .4649 = .0351. The probability that a house is listed for more than $350 000 is 3.51%.

2. We will require two z-values corresponding to list prices of $150 000 and $300 000. Note that the values are on different sides of the mean. From formula (6–1):

$$z = \frac{X - \mu}{\sigma} = \frac{150\,000 - 192\,712}{86\,881} \quad \text{and} \quad z = \frac{X - \mu}{\sigma} = \frac{300\,000 - 192\,712}{86\,881}$$
$$= -.49 \quad\quad\quad\quad\quad\quad\quad\quad\quad\quad\quad\quad = 1.23$$

See Appendix D. Move down the left margin to the row headed .4 and across that row to the column headed 0.09. The value is .1879. Then, move down the left margin to the row headed 1.2 and across that row to the column headed 0.03. The value is .3907. However, the required area is between $150 000 and $300 000, so we must add the two areas. So, .1879 + .3907 = .5786. The probability that a house is listed for between $150 000 and $300 000 is 57.86%.

3. Again, we will require two z-values corresponding to list prices of $75 000 and $125 000. Note that the values are on the same side of the mean. From formula (6–1):

$$z = \frac{X - \mu}{\sigma} = \frac{75\,000 - 192\,712}{86\,881} \quad \text{and} \quad z = \frac{X - \mu}{\sigma} = \frac{125\,000 - 192\,712}{86\,881}$$
$$= -1.35 \quad\quad\quad\quad\quad\quad\quad\quad\quad\quad\quad\quad = -.78$$

See Appendix D. Move down the left margin to the row headed 1.3 and across that row to the column headed 0.05. The value is .4115. Then, move down the left margin to the row headed .7 and across that row to the column headed 0.08. The value is .2823. However, the required area is between $75 000 and $125 000, so we must subtract the two areas. So, .4115 − .2823 = .1292. The probability that a house is listed for between $75 000 and $125 000 is 12.92%.

MegaStat produces the following output.

1.

p(lower)	p(upper)	z	x	mean	std. dev
.9649	.0351	1.81	350 000	192712	86 881

The required area is p(upper) = 3.51%.

2.

p(lower)	p(upper)	z	x	mean	std. dev
.3115	.6885	−0.49	150 000	192 712	86 881

p(lower)	p(upper)	z	x	mean	std. dev
.8916	.1084	1.23	300 000	192 712	86 881

The p(lower) of .3115 is the area under the normal curve up to $150 000. We do not need this area. The p(lower) of .8916 is the area under the normal curve up to $300 000. So, we subtract the area we do not need, .8916 − .3115 = .5801 (58.01%). The probability is slightly different than that calculated using tables due to rounding.

3.

p(lower)	p(upper)	z	x	mean	std. dev
.0877	.9123	−1.35	75 000	192 712	86 881

p(lower)	p(upper)	z	x	mean	std. dev
.2179	.7821	−0.78	125 000	192 712	86 881

This is calculated in the same manner as #2. The p(lower) of .0877 is the area under the normal curve up to $75 000. We do not need this area. The p(lower) of .2179 is the area under the normal curve up to $125 000. So, we subtract the area we do not need, .2179 − .0877 = .1302 (13.02%). The probability is slightly different than that calculated using tables due to rounding.

Note: Further examples and instructions on using Excel and MegaStat are in the appendix section on the CD-ROM.

Self-Review 6–4

Refer to the previous example, where the distribution of weekly incomes follows the normal distribution with a mean of $1000 and the standard deviation is $100.

(a) What percent of the shift supervisors earn a weekly income between $750 and $1225? Draw a normal curve and shade the desired area on your diagram.
(b) What percent of the shift supervisors earn a weekly income between $1100 and $1225? Draw a normal curve and shade the desired area on your diagram.

Exercises

11. A normal distribution has a mean of 50 and a standard deviation of 4.
    a. Compute the probability of a value between 44.0 and 55.0.
    b. Compute the probability of a value greater than 55.0.
    c. Compute the probability of a value between 52.0 and 55.0.
12. A normal population has a mean of 80.0 and a standard deviation of 14.0.
    a. Compute the probability of a value between 75.0 and 90.0.
    b. Compute the probability of a value 75.0 or less.
    c. Compute the probability of a value between 55.0 and 70.0.
13. A cola-dispensing machine is set to dispense on average 225 ml of cola per cup. The standard deviation is 10 ml. The distribution amounts dispensed follows a normal distribution.
    a. What is the probability that the machine will dispense between 235 ml and 250 ml of cola?
    b. What is the probability that the machine will dispense 250 ml of cola or more?
    c. What is the probability that the machine will dispense between 205 ml and 250 ml of cola?

The Normal Probability Distribution

14. The amounts of money requested on home loan applications at Down River Federal Savings follow the normal distribution, with a mean of $70 000 and a standard deviation of $20 000. A loan application is received this morning. What is the probability:
    a. The amount requested is $80 000 or more?
    b. The amount requested is between $65 000 and $80 000?
    c. The amount requested is $65 000 or more?
15. WNAE, an all-news AM station, finds that the distribution of the lengths of time listeners are tuned to the station follows the normal distribution. The mean of the distribution is 15.0 minutes and the standard deviation is 3.5 minutes. What is the probability that a particular listener will tune in:
    a. More than 20 minutes?
    b. For 20 minutes or less?
    c. Between 10 and 12 minutes?
16. The mean starting salary for college graduates in the spring of 2002 was $31 280. Assume that the distribution of starting salaries follows the normal distribution with a standard deviation of $3300. What percent of the graduates have starting salaries:
    a. Between $30 000 and $35 000?
    b. More than $40 000?
    c. Between $35 000 and $40 000?

Previous examples require finding the percent of the observations located between two observations or the percent of the observations above, or below, a particular observation X. A further application of the normal distribution involves finding the value of the observation X when the percent above or below the observation is given.

EXAMPLE

The Layton Tire and Rubber Company wishes to set a minimum distance guarantee on its new MX100 tire. Tests reveal the mean number of kilometres is 109 000 with a standard deviation of 3300 km and that the distribution of kilometres follows the normal distribution. They want to set the minimum guaranteed number of kilometres so that no more than 4 percent of the tires will have to be replaced. What minimum guaranteed kilometres should Layton announce?

Solution

The details of this case are shown in the following diagram, where X represents the minimum guaranteed number of kilometres.

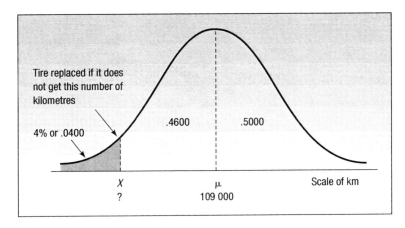

Inserting these values in formula (6–1) for z:

$$z = \frac{X - \mu}{\sigma} = \frac{X - 109\,000}{3300}$$

Notice that there are two unknowns, z and X. To find X, we first find z, and then solve for X. Notice the area under the normal curve to the left of μ is .5000. The area between μ and X is .4600, found by .5000 − .0400. Now refer to Appendix D. Search the body of the table for the area closest to .4600. The closest area is .4599. Move to the margins from this value and read the z value of 1.75. Because the value is to the left of the mean, it is actually −1.75. These steps are illustrated in Table 6–2.

TABLE 6–2 Selected Areas under the Normal Curve

z	.03	.04	.05	.06
⋮				
1.5	.4370	.4382	.4394	.4406
1.6	.4484	.4495	.4505	.4515
1.7	.4582	.4591	.4599	.4608
1.8	.4664	.4671	.4678	.4686

Knowing that the distance between μ and X is -1.75σ or $z = -1.75$, we can now solve for X (the minimum guaranteed kilometres):

$$z = \frac{X - 109\,000}{3300}$$

$$-1.75 = \frac{X - 109\,000}{3300}$$

$$-1.75(3300) = X - 109\,000$$

$$X = 109\,000 - 1.75(3300) = 103\,225$$

So Layton can advertise that it will replace for free any tire that wears out before it reaches 103 225 km, and the company will know that only 4 percent of the tires will be replaced under this plan.

Excel and MegaStat will also find the number of kilometres. Excel commands follow.

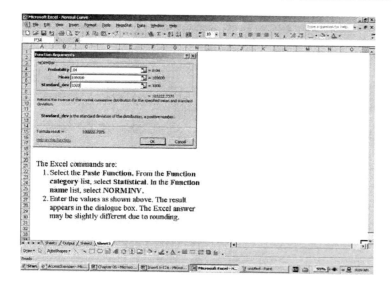

The Excel commands are:
1. Select the **Paste Function**. From the **Function category** list, select **Statistical**. In the **Function name** list, select **NORM.INV**.
2. Enter the values as shown above. The result appears in the dialogue box. The Excel answer may be slightly different due to rounding.

MegaStat commands follow.

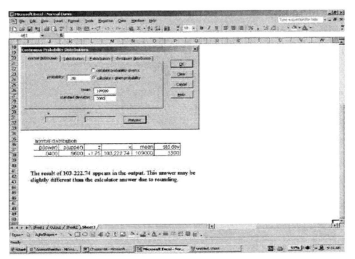

Self-Review 6–5

An analysis of the final test scores for Introduction to Business reveals the scores follow the normal distribution. The mean of the distribution is 75 and the standard deviation is 8. The professor wants to award an A to students whose score is in the highest 10 percent. What is the dividing point for those students who earn an A and those earning a B?

Exercises

17. A normal distribution has a mean of 50 and a standard deviation of 4. Determine the value below which 95 percent of the observations will occur.
18. A normal distribution has a mean of 80 and a standard deviation of 14. Determine the value above which 80 percent of the values will occur.
19. The amounts dispensed by a cola machine follow the normal distribution with a mean of 200 ml and a standard deviation of 0.3 ml per cup. How much cola is dispensed in the largest 1 percent of the cups?
20. Refer to Exercise 14, where the amount requested for home loans followed the normal distribution with a mean of $70 000 and a standard deviation of $20 000.
    a. How much is requested on the largest 3 percent of the loans?
    b. How much is requested on the smallest 10 percent of the loans?
21. Assume that the mean hourly cost to operate a commercial airplane follows the normal distribution with a mean $2100 per hour and a standard deviation of $250. What is the operating cost for the lowest 3 percent of the airplanes?
22. The monthly sales of mufflers follow the normal distribution with a mean of 1200 and a standard deviation of 225. The manufacturer would like to establish inventory levels such that there is only a 5 percent chance of running out of stock. Where should the manufacturer set the inventory levels?

The Normal Approximation to the Binomial

Chapter 5 describes the binomial probability distribution, which is a discrete distribution. The table of binomial probabilities in Appendix A goes successively from an n of 1 to an n of 20, and then to $n = 25$. If a problem involved taking a sample of 60, generating a binomial distribution for that large a number would be very time consuming. A more efficient approach is to apply the *normal approximation to the binomial*.

184 Chapter 6

Using the normal distribution (a continuous distribution) as a substitute for a binomial distribution (a discrete distribution) for large values of n seems reasonable because, as n increases, a binomial distribution gets closer and closer to a normal distribution. Chart 6–5 depicts the change in the shape of a binomial distribution with $p = .50$ from an n of 1, to an n of 3, to an n of 20. Notice how the case where $n = 20$ approximates the shape of the normal distribution. That is, compare the case where $n = 20$ to the normal curve in Chart 6–1 on page 184.

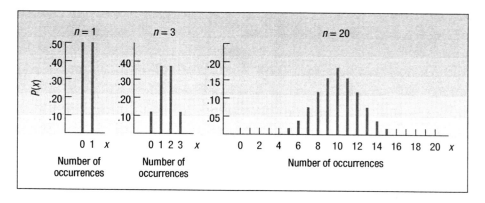

CHART 6–5 Binomial Distributions for an n of 1, 3, and 20, Where $p = .50$

When to use the normal approximation

When can we use the normal approximation to the binomial? The normal probability distribution is a good approximation to the binomial probability distribution when np and $n(1 - p)$ are both at least 5. However, before we apply the normal approximation, we must make sure that our distribution of interest is in fact a binomial distribution. Recall from Chapter 5 that four criteria must be met:

1. There are only two mutually exclusive outcomes to an experiment: a "success" and a "failure."
2. The distribution results from counting the number of successes in a fixed number of trials.
3. Each trial is independent.
4. The probability, p, remains the same from trial to trial.

Continuity Correction Factor

To show the application of the normal approximation to the binomial and the need for a correction factor, suppose the management of the Santoni Pizza Restaurant found that 70 percent of their new customers return for another meal. For a week in which 80 new (first-time) customers dined at Santoni's, what is the probability that 60 or more will return for another meal?

Notice the binomial conditions are met: (1) There are only two possible outcomes—a customer either returns for another meal or does not return. (2) We can count the number of successes, meaning, for example, that 57 of the 80 customers return. (3) The trials are independent, meaning that if the 34th person returns for a second meal, that does not affect whether the 58th person returns. (4) The probability of a customer returning remains at .70 for all 80 customers.

Therefore, we could use the binomial formula (5–3) described on page 159.

$$P(x) = {}_nC_x\,(p)^x\,(1 - p)^{n-x}$$

To find the probability 60 or more customers return for another pizza, we need to first find the probability exactly 60 customers return. That is:

$$P(x = 60) = {}_{80}C_{60}\,(.70)^{60}\,(1 - .70)^{20} = .063$$

The Normal Probability Distribution

Next we find the probability that exactly 61 customers return. It is:

$$P(x = 61) = {}_{80}C_{61} (.70)^{61} (1 - .70)^{19} = .048$$

We continue this process until we have the probability all 80 customers return. Finally, we add the probabilities from 60 to 80. Solving the above problem in this manner is tedious. We can also use a computer software package such as MINITAB or Excel to find the various probabilities. Listed on the next page are the binomial probabilities for $n = 80$, $p = .70$, and x, the number of customers returning, ranging from 43 to 68. The probability of any number of customers less than 43 or more than 68 returning is less than .001.

Number Returning	Probability	Number Returning	Probability
43	0.001	56	0.097
44	0.002	57	0.095
45	0.003	58	0.088
46	0.006	59	0.077
47	0.009	60	0.063
48	0.015	61	0.048
49	0.023	62	0.034
50	0.033	63	0.023
51	0.045	64	0.014
52	0.059	65	0.008
53	0.072	66	0.004
54	0.084	67	0.002
55	0.093	68	0.001

We can find the probability of 60 or more returning by summing $0.063 + 0.048 + \cdots + 0.001$, which is 0.197. However, a look at the plot below shows the similarity of this distribution to a normal distribution. All we need do is "smooth out" the discrete probabilities into a continuous distribution. Furthermore, working with a normal distribution will involve far fewer calculations than working with the binomial.

The trick is to let the discrete probability for 56 customers be represented by an area under the continuous curve between 55.5 and 56.5. Then let the probability for 57 customers be represented by an area between 56.5 and 57.5 and so on. This is just the opposite of rounding off the numbers to a whole number.

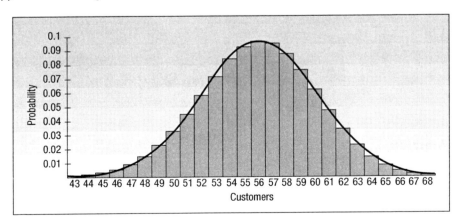

Because we use the normal distribution to determine the binomial probability of 60 or more successes, we must subtract, in this case, .5 from 60. The value .5 is called the **continuity correction factor.** This small adjustment must be made because a continuous distribution (the normal distribution) is being used to approximate a discrete distribution (the binomial distribution). Subtracting, $60 - .5 = 59.5$.

Chapter 6

CONTINUITY CORRECTION FACTOR The value .5 subtracted or added, depending on the question, to a selected value when a discrete probability distribution is approximated by a continuous probability distribution.

Statistics in Action

The heights of adults approximate a normal distribution, but those individuals who are very tall can cause design problems. For example, Shaquille O'Neal, a professional basketball player with the Los Angeles Lakers, is 218 cm tall. The height of the standard doorway is 203 cm, so Shaquille and many other NBA players duck to get into most rooms.

As another example, the driver's seat in most vehicles is set to comfortably fit a person who is at least 159 cm tall. The distribution of heights of adult women is approximately a normal distribution with a mean of 161.5 cm and a standard deviation of 6.3 cm. Thus about 35 percent of adult women will not fit comfortably in the driver's seat.

How to Apply the Correction Factor

Only four cases may arise. These cases are:

1. For the probability that *at least X* occur, use the area *above* $(X - .5)$.
2. For the probability that *more than X* occur, use the area *above* $(X + .5)$.
3. For the probability that *X or fewer* occur, use the area *below* $(X + .5)$.
4. For the probability that *fewer than X* occur, use the area *below* $(X - .5)$.

To use the normal distribution to approximate the probability that 60 or more first-time Santoni customers out of 80 will return, follow the procedure shown below.

Step 1. Find the z corresponding to an X of 59.5 using formula (6–1), and formulas (5–4) and (5–5) for the mean and the variance of a binomial distribution:

$$\mu = np = 80(.70) = 56$$
$$\sigma^2 = np(1 - p) = 80(.70)(1 - .70) = 16.8$$
$$\sigma = \sqrt{16.8} = 4.10$$
$$z = \frac{X - \mu}{\sigma} = \frac{59.5 - 56}{4.10} = 0.85$$

Step 2. Determine the area under the normal curve between a μ of 56 and an X of 59.5. From step 1, we know that the z value corresponding to 59.5 is 0.85. So we go to Appendix D and read down the left margin to 0.8, and then we go horizontally to the area under the column headed by .05. That area is .3023.

Step 3. Calculate the area beyond 59.5 by subtracting .3023 from .5000 (.5000 − .3023 = .1977). Thus, .1977 is the probability that 60 or more first-time Santoni customers out of 80 will return for another meal. In probability notation P(customers > 59.5) = .5000 − .3023 = .1977. The details of this problem are shown graphically:

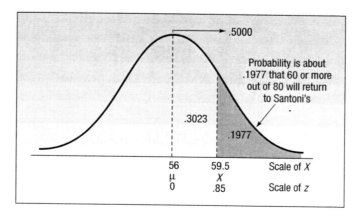

No doubt you will agree that using the normal approximation to the binomial is a more efficient method of estimating the probability of 60 or more first-time customers returning. The result compares favourably with that computed on page 209, using the binomial distribution. The probability using the binomial distribution is .197, whereas the probability using the normal approximation is .1977.

The Normal Probability Distribution

Self-Review 6–6

A study by Great Southern Home Insurance revealed that none of the stolen goods were recovered by homeowners in 80 percent of reported thefts.

(a) During a period in which 200 thefts occurred, what is the probability that no stolen goods were recovered in 170 or more of the robberies?
(b) During a period in which 200 thefts occurred, what is the probability that no stolen goods were recovered in 150 or more robberies?

Exercises

23. Assume a binomial probability distribution with $n = 50$ and $p = .25$. Compute the following:
    a. The mean and standard deviation of the random variable.
    b. The probability that X is 15 or more.
    c. The probability that X is 10 or less.
24. Assume a binomial probability distribution with $n = 40$ and $p = .55$. Compute the following:
    a. The mean and standard deviation of the random variable.
    b. The probability that X is 25 or greater.
    c. The probability that X is 15 or less.
    d. The probability that X is between 15 and 25 inclusive.
25. Dottie's Tax Service specializes in tax returns for professional clients, such as physicians, dentists, accountants, and lawyers. A recent audit of the returns she prepared indicated that an error was made on 5 percent of the returns she prepared last year. Assuming this rate continues into this year and she prepares 60 returns, what is the probability that she makes errors on:
    a. More than six returns?
    b. At least six returns?
    c. Exactly six returns?
26. Shorty's Muffler advertises they can install a new muffler in 30 minutes or less. However, the work standards department at corporate headquarters recently conducted a study and found that 20 percent of the mufflers were not installed in 30 minutes or less. The Maumee branch installed 50 mufflers last month. If the corporate report is correct:
    a. How many of the installations at the Maumee branch would you expect to take more than 30 minutes?
    b. What is the likelihood that fewer than eight installations took more than 30 minutes?
    c. What is the likelihood that eight or fewer installations took more than 30 minutes?
    d. What is the likelihood that exactly 8 of the 50 installations took more than 30 minutes?
27. A study conducted by the nationally known Taurus Health Club revealed that 30 percent of its new members are significantly overweight. A membership drive in a metropolitan area resulted in 500 new members.
    a. It has been suggested that the normal approximation to the binomial be used to determine the probability that 175 or more of the new members are significantly overweight. Does this problem qualify as a binomial problem? Explain.
    b. What is the probability that 175 or more of the new members are significantly overweight?
    c. What is the probability that 140 or more new members are significantly overweight?
28. Research on new juvenile delinquents put on probation revealed that 38 percent of them committed another crime.
    a. What is the probability that of the last 100 new juvenile delinquents put on probation, 30 or more will commit another crime?
    b. What is the probability that 40 or fewer of the delinquents will commit another crime?
    c. What is the probability that from 30 to 40 of the delinquents will commit another crime?

Chapter Outline

I. The normal distribution is a continuous probability distribution with the following characteristics.
   A. It is bell-shaped and the mean, median, and mode are equal.
   B. It is symmetrical.

Chapter 6

C. It is asymptotic, meaning the curve approaches but never touches the X-axis.
D. It is completely described by the mean and standard deviation.
E. There is a family of normal distributions. Each time the mean or standard deviation changes, a new distribution is created.
F. The area under a normal curve expresses the probability of an outcome.

II. The standard normal distribution is a particular normal distribution.
  A. It has a mean of 0 and a standard deviation of 1.
  B. Any normal distribution can be converted to the standard normal distribution by the following formula.

$$z = \frac{X - \mu}{\sigma} \qquad [6\text{-}1]$$

  C. By standardizing a normal distribution, we can report the distance from the mean in units of the standard deviation.

III. The normal distribution can approximate a binomial distribution under certain conditions.
  A. np and $n(1 - p)$ must both be at least 5.
    1. n is the number of observations.
    2. p is the probability of a success.
  B. The four conditions for a binomial distribution are:
    1. There are only two possible outcomes.
    2. p remains the same from trial to trial.
    3. The trials are independent.
    4. The distribution results from a count of the number of successes in a fixed number of trials.
  C. The mean and variance of a binomial distribution are computed as follows:

$$\mu = np$$
$$\sigma^2 = np(1 - p)$$

  D. The continuity correction factor of .5 is used to extend the continuous value of X one-half unit in either direction. This correction compensates for estimating a discrete distribution by a continuous distribution.

Chapter Exercises

29. The net sales and the number of employees for aluminum fabricators with similar characteristics are organized into frequency distributions. Both are normally distributed. For the net sales, the mean is $180 million and the standard deviation is $25 million. For the number of employees, the mean is 1500 and the standard deviation is 120. Clarion Fabricators had sales of $170 million and 1850 employees.
    a. Convert Clarion's sales and number of employees to z values.
    b. Locate the two z values.
    c. Compare Clarion's sales and number of employees with those of the other fabricators.

30. The accounting department at Weston Materials, Inc., a national manufacturer of unattached garages, reports that it takes two construction workers a mean of 32 hours and a standard deviation of 2 hours to erect the Red Barn model. Assume the assembly times follow the normal distribution.
    a. Determine the z values for 29 and 34 hours. What percent of the garages take between 32 hours and 34 hours to erect?
    b. What percent of the garages take between 29 hours and 34 hours to erect?
    c. What percent of the garages take 28.7 hours or less to erect?
    d. Of the garages, 5 percent take how many hours or more to erect?

31. A recent report indicated a typical family of four spends $490 per month on food. Assume the distribution of food expenditures for a family of four follows the normal distribution, with a mean of $490 and a standard deviation of $90.
    a. What percent of the families spend more than $30 but less than $490 per month on food?
    b. What percent of the families spend less than $430 per month on food?

The Normal Probability Distribution

　　c. What percent spend between $430 and $600 per month on food?
　　d. What percent spend between $500 and $600 per month on food?

32. A study of long distance phone calls made from the corporate offices of the Pepsi Bottling Group Inc. showed the calls follow the normal distribution. The mean length of time per call was 4.2 minutes and the standard deviation was 0.60 minutes.
    a. What fraction of the calls last between 4.2 and 5 minutes?
    b. What fraction of the calls last more than 5 minutes?
    c. What fraction of the calls last between 5 and 6 minutes?
    d. What fraction of the calls last between 4 and 6 minutes?
    e. As part of her report to the president the Director of Communications would like to report the length of the longest (in duration) 4 percent of the calls. What is this time?

33. Shaver Manufacturing, Inc. offers dental insurance to its employees. A recent study by the Human Resource Director shows the annual cost per employee per year followed the normal distribution, with a mean of $1280 and a standard deviation of $420 per year.
    a. What fraction of the employees cost more than $1500 per year for dental expenses?
    b. What fraction of the employees cost between $1500 and $2000 per year?
    c. Estimate the percent that did not have any dental expense.
    d. What was the cost for the 10 percent of employees that incurred the highest dental expense?

34. The annual commissions earned by sales representatives of Machine Products Inc., a manufacturer of light machinery, follow the normal distribution. The mean yearly amount earned is $40 000 and the standard deviation is $5000.
    a. What percent of the sales representatives earn more than $42 000 per year?
    b. What percent of the sales representatives earn between $32 000 and $42 000?
    c. What percent of the sales representatives earn between $32 000 and $35 000?
    d. The sales manager wants to award the sales representatives who earn the largest commissions a bonus of $1000. He can award a bonus to 20 percent of the representatives. What is the cutoff point between those who earn a bonus and those who do not?

35. The masses of cans of Monarch pears follow the normal distribution with a mean of 1000 g and a standard deviation of 50 g. Calculate the percentage of the cans that have masses:
    a. Less than 860 g.
    b. Between 1055 g and 1100 g.
    c. Between 860 g and 1055 g.

36. The number of passengers on the *Queen Elizabeth II* during one-week cruises in the Caribbean follows the normal distribution. The mean number of passengers per cruise is 1820 and the standard deviation is 120.
    a. What percent of the cruises will have between 1820 and 1970 passengers?
    b. What percent of the cruises will have 1970 passengers or more?
    c. What percent of the cruises will have 1600 or fewer passengers?
    d. How many passengers are on the cruises with the fewest 25 percent of passengers?

37. Management at Gordon Electronics is considering adopting a bonus system to increase production. One suggestion is to pay a bonus on the highest 5 percent of production based on past experience. Past records indicate weekly production follows the normal distribution. The mean of this distribution is 4000 units per week and the standard deviation is 60 units per week. If the bonus is paid on the upper 5 percent of production, the bonus will be paid on how many units or more?

38. Fast Service Truck Lines uses the Ford Super 1310 exclusively. Management made a study of the maintenance costs and determined the number of kilometres traveled during the year followed the normal distribution. The mean of the distribution was 96 600 km and the standard deviation 3200 km.
    a. What percent of the Ford Super 1310s logged 104 970 km or more?
    b. What percent of the trucks logged more than 91 800 but less than 93 800 km?
    c. What percent of the Fords traveled 99 800 or less during the year?
    d. Is it reasonable to conclude that any of the trucks were driven more than 112 700 km? Explain.

39. Best Electronics offers a "no hassle" returns policy. The number of items returned per day follows the normal distribution. The mean number of customer returns is 10.3 per day and the standard deviation is 2.25 per day.
    a. In what percent of the days are there 8 or fewer customers returning items?
    b. In what percent of the days are between 12 and 14 customers returning items?
    c. Is there any chance of a day with no returns?

www.mcgrawhill.ca/college/lind

Chapter 6

40. A recent study shows that 20 percent of all employees prefer their vacation time during March break. If a company employs 50 people, what is the probability that:
    a. Fewer than 5 employees want their vacation during March break?
    b. More than 5 employees want their vacation during March break?
    c. Exactly 5 employees want their vacation during March break?
    d. More than 5 but fewer than 15 employees want their vacation during March break?

41. According to Statistics Canada, the mean number of hours of TV viewing per week is higher among adult women than men. A recent study showed women spent an average of 26.3 hours per week watching TV and men 20.7 per week. (Statistics Canada, CANSIM tables 502-0002 and 502-0003; retrieved 14/04/2005; www.statcan.ca/english/Pgdb/arts23.htm). Assume that the distribution of hours watched follows a normal distribution for both groups, and that the standard deviations are 4.5 hours for the women and 5.1 hours for the men.
    a. What percent of women watch TV less than 30 hours per week?
    b. What percent of men watch TV more than 18 hours per week?
    c. How many hours of TV do the one percent of women who watch the most TV per week watch? Find the comparable value for the men.

42. The current model Boeing 737 has a capacity of 189 passengers. The distribution of the number of seats sold follows a normal distribution with a mean of 155 seats and a standard deviation of 15 seats.
    a. What is the probability of selling more than 134 seats?
    b. What is the probability of selling less than 173 seats?
    c. What is the probability of selling between 134 and 173 seats?
    d. What percent of the time does an airline sell more seats than there are available?

43. The median age in Canada in 2003 was 37.9 years (Statistics Canada, CANSIM database table 051-0001; retrieved 03/03/2005; http://cansim2.statcan.ca). Assume that the distribution of median ages is normally distributed with a standard deviation of 1.4 years.
    a. What is the probability that the median age is more than 37.0 years?
    b. What is the probability that the median age is less than 39.5 years?
    c. What is the probability that the median age is between 36.0 and 37.0 years?
    d. What is the probability that the median age is less than 36.0 or more than 37.0 years?

44. The funds dispensed at the ATM at the Queensway site of the Trillium Hospital follows a normal distribution with a mean of $4200 per day and a standard deviation of $720 per day. The machine is programmed to notify the bank if the amount dispensed is very low, less than $2500, or very high, more than $6000.
    a. What percent of the time will the bank be notified because the amount dispensed is very low?
    b. What percent of the time will the bank be notified because the amount dispensed is very high?
    c. What percent of the time will the bank not be notified regarding the amount of funds dispensed?

45. A recent survey reported that 64 percent of men over the age of 18 consider nutrition a top priority in their lives. Suppose we select a sample of 60 men. What is the likelihood that:
    a. 32 or more consider nutrition important?
    b. 44 or more consider nutrition important?
    c. More than 32 but fewer than 43 consider nutrition important?
    d. Exactly 44 consider diet important?

46. It is estimated that 10 percent of those taking the quantitative methods portion of the CPA examination fail that section. Sixty students are taking the exam this Saturday.
    a. How many would you expect to fail? What is the standard deviation?
    b. What is the probability that exactly two students will fail?
    c. What is the probability at least two students will fail?

47. The Woodbridge Traffic Division reported 40 percent of the high-speed chases involving automobiles result in a minor or major accident. During a month in which 50 high-speed chases occur, what is the probability that 25 or more will result in a minor or major accident?

48. Cruise ships of the Royal Viking line report that 80 percent of their rooms are occupied during September. For a cruise ship having 800 rooms, what is the probability that 665 or more are occupied in September?

49. The goal at airports handling international flights is to clear these flights within 45 minutes. Let's interpret this to mean that 95 percent of the flights are cleared in 45 minutes, so 5 percent of the flights take longer to clear. Let's also assume that the distribution is approximately normal.

The Normal Probability Distribution

a. If the standard deviation of the time to clear an international flight is 5 minutes, what is the mean time to clear a flight?
b. Suppose the standard deviation is 10 minutes, not the 5 minutes suggested in part a. What is the new mean?
c. A customer has 30 minutes from the time her flight landed to catch her limousine. Assuming a standard deviation of 10 minutes, what is the likelihood that she will be cleared in time?

50. An Air Force study indicates that the probability of a disaster such as the January 28, 1986, explosion of the space shuttle *Challenger* was 1 in 35. The *Challenger* flight was the 25th mission.
    a. How many disasters would you expect in the first 25 flights?
    b. Evaluate the normal approximation to estimate the probability of at least one disaster in 25 missions. Is this a good approximation? Tell why or why not.

51. The registrar at Elmwood University studied the grade point averages (GPAs) of students over many years. Assume the GPA distribution follows a normal distribution with a mean of 3.10 and a standard deviation of 0.30.
    a. What is the probability that a randomly selected Elmwood student has a GPA between 2.00 and 3.00?
    b. What percent of the students are on probation, that is, have a GPA less than 2.00?
    c. The student population at EU is 10 000. How many students are on the dean's list, that is, have GPAs of 3.70 or higher?
    d. To qualify for a Bell scholarship, a student must be in the top 10 percent. What GPA must a student attain to qualify for a Bell scholarship?

52. Jon Molnar will graduate from Cobden High School this year. He took the American College Test (ACT) for college admission and received a score of 30. The high school principal informed him that only 2 percent of the students taking the exam receive a higher score. The mean score for all students taking the exam is 18.3. Jon's friends Karrie and George also took the test but were not given any information by the principal other than their scores. Karrie scored 25 and George 18. Based on this information, what were Karrie's and George's percentile ranks? Assume that the distribution of scores follows the normal distribution.

53. The weights of canned hams processed at the Henline Ham Company follow the normal distribution, with a mean of 3.1 kg and a standard deviation of .125 kg. The label weight is given as 3 kg.
    a. What proportion of the hams actually weigh less than the amount claimed on the label?
    b. The owner, Glen Henline, is considering two proposals to reduce the proportion of hams below label weight. He can increase the mean weight to 3.15 kg and leave the standard deviation the same, or he can leave the mean weight at 3.1 kg and reduce the standard deviation from .125 kg to .10 kg. Which change would you recommend?

54. The Gravenhurst *Enquirer* reported that the mean number of hours worked per week by those employed full time is 43.9. The article further indicated that about one third of those employed full time work less than 40 hours per week.
    a. Given this information and assuming that number of hours worked follows the normal distribution, what is the standard deviation of the number of hours worked?
    b. The article also indicated that 20 percent of those working full time work more than 49 hours per week. Determine the standard deviation with this information. Are the two estimates of the standard deviation similar? What would you conclude?

55. Most four-year automobile leases allow up to 100 000 km. If the lessee goes beyond this amount, a penalty of 8 cents per kilometre is added to the lease cost. Suppose the distribution of kilometres driven on four-year leases follows the normal distribution. The mean is 85 000 kilometres and the standard deviation is 8000 kilometres.
    a. What percent of the leases will yield a penalty because of excess distance traveled?
    b. If the automobile company wanted to change the terms of the lease so that 25 percent of the leases went over the limit, where should the new upper limit be set?
    c. One definition of a low-mileage car is one that is 4 years old and has been driven less than 72 000 kilometres. What percent of the cars returned are considered low-mileage?

56. The price of shares of the Continental Bank at the end of trading each day for the last year followed the normal distribution. Assume there were 240 trading days in the year. The mean price was $42.00 per share and the standard deviation was $2.25 per share.
    a. What percent of the days was the price over $45.00? How many days would you estimate?
    b. What percent of the days was the price between $38.00 and $40.00?
    c. What was the stock's price on the *highest* 15 days of the year?

Chapter 6

57. The annual sales of romance novels follows the normal distribution. However, the mean and the standard deviation are unknown. Forty percent of the time sales are more than 470 000, and 10 percent of the time sales are more than 500 000. What are the mean and the standard deviation?

58. In establishing warranties on HDTV sets, the manufacturer wants to set the limits so that few will need repair at manufacturer expense. On the other hand, the warranty period must be long enough to make the purchase attractive to the buyer. For a new HDTV the mean number of months until repairs are needed is 36.84 with a standard deviation of 3.34 months. Where should the warranty limits be set so that only 10 percent of the HDTVs need repairs at the manufacturer's expense?

59. DeKorte Tele-Marketing Inc. is considering purchasing a machine that randomly selects and automatically dials telephone numbers. DeKorte Tele-Marketing makes most of its calls during the evening, so calls to business phones are wasted. The manufacturer of the machine claims that their programming reduces the calling to business phones to 15 percent of all calls. To test this claim the Director of Purchasing at DeKorte programmed the machine to select a sample of 150 phone numbers. What is the likelihood that more than 30 of the phone numbers selected are that of a business, assuming the manufacturer's claim is correct?

Data Set Exercises

60. Refer to the Real Estate data, Regina & Surrounding Area, on the CD-ROM, which reports information on listed homes and townhomes, March 2005. The mean list price (in $ thousands) of the homes was computed earlier. Use the normal distribution to estimate the percent of homes listed for more than $280 000. Compare this to the actual results. Does the normal distribution yield a good approximation of the actual results?

61. Refer to the CREA (Canadian Real Estate Association) data on the CD-ROM, which reports information on average house prices nationally and in a selection of cities across Canada for January and March, 2004 and 2005. Select the cities only.
    a. Calculate the mean and the standard deviation for Jan 05. Use the normal distribution to estimate the number of homes listed for less than $200 000. Compare the estimate to the actual results. Does the normal distribution yield a good approximation of the actual results?
    b. Calculate the mean and the standard deviation for Jan 04. Use the normal distribution to estimate the number of homes listed for more than $300 000. Compare the estimate to the actual results. Does the normal distribution yield a good approximation of the actual results?

62. Refer to the International data, which reports demographic and economic information on 46 countries.
    a. The mean of the GDP/capita variable is 16.58 with a standard deviation of 9.27. Use the normal distribution to estimate the percentage of countries with exports above 24. Compare this estimate with the actual proportion. Does the normal distribution appear accurate in this case? Explain.
    b. The mean of the Exports is 116.3 with a standard deviation of 157.4. Use the normal distribution to estimate the percentage of countries with Exports above 170. Compare this estimate with the actual proportion. Does the normal distribution appear accurate in this case? Explain.

Additional exercises that require you to access information at related Internet sites are available on the CD-ROM included with this text.

The Normal Probability Distribution

Chapter 6 Answers to Self-Reviews

6–1 (a) 2.25, found by:
$$z = \frac{\$1225 - \$1000}{\$100} = \frac{\$225}{\$100} = 2.25$$

(b) −2.25, found by:
$$z = \frac{\$775 - \$1000}{\$100} = \frac{-\$225}{\$100} = -2.25$$

6–2 (a) $46 400 and $48 000, found by $47 200 ± 1($800).
(b) $45 600 and $48 800, found by $47 200 ± 2($800).
(c) $44 800 and $49 600, found by $47 200 ± 3($800).
(d) $47 200. Mean, median, and mode are equal for a normal distribution.
(e) Yes, a normal distribution is symmetrical.

6–3 (a) Computing z:
$$z = \frac{482 - 400}{50} = +1.64$$

Referring to Appendix D, the area is .4495.
P(400 < rating < 482) = .4495

(b) .0505, found by .5000 − .4495.
P(rating > 482) = .5000 − .4495 = .0505

(c)
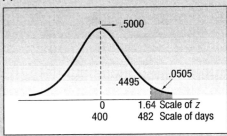

6–4 (a) 98.16%, found by 0.4938 + 0.4878.

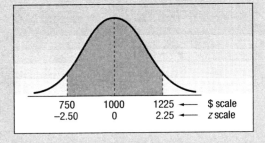

(b) 14.65%, found by 0.4878 − 0.3413.

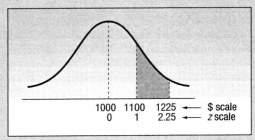

6–5 85.24 (instructor would no doubt make it 85). The closest area to .4000 is .3997; z is 1.28. Then:
$$1.28 = \frac{X - 75}{8}$$
$$10.24 = X - 75$$
$$X = 85.24$$

6–6 (a) .0465, found by $\mu = np = 200(.80) = 160$, and $\sigma^2 = np(1-p) = 200(.80)(1-.80) = 32$. Then,
$$\sigma = \sqrt{32} = 5.66$$
$$z = \frac{169.5 - 160}{5.66} = 1.68$$

Area from Appendix D is .4535. Subtracting from .5000 gives .0465.

(b) .9686, found by .4686 + .5000. First calculate z:
$$z = \frac{149.5 - 160}{5.66} = -1.86$$

Area from Appendix D is .4686.